57914

RECENT ADVANCES IN
ANIMAL NUTRITION — 1997

Recent Advances in Animal Nutrition

1997

P.C. Garnsworthy, PhD
J. Wiseman, PhD
University of Nottingham

NOTTINGHAM
University Press

Nottingham University Press
Manor Farm, Main Street, Thrumpton
Nottingham, NG11 0AX, United Kingdom

NOTTINGHAM

First published 1997

British Library Cataloguing in Publication Data
Recent Advances in Animal Nutrition — 1997:
University of Nottingham Feed Manufacturers
Conference (31st, 1997, Nottingham)
I. Garnsworthy, Philip C. II. Wiseman, J.

ISBN 1-897676-042

Typeset by Nottingham University Press, Nottingham
Printed and bound by Redwood Books, Trowbridge, Wiltshire

PREFACE

This book contains the proceedings of the 31st University of Nottingham Feed Manufacturers Conference, held at Sutton Bonington in January 1997. Due to the short period of time between the New Year holiday and the start of the University term this year, the length of the conference had to be reduced to two days. Consequently, fewer papers were presented than in previous years, but this did not affect their quality. The conference was arranged into three sessions on General Nutrition, Poultry Nutrition and Ruminant Nutrition. Although there was no session specifically dedicated to pigs this year, the papers in the General session are heavily slanted towards this species.

The three papers in the General Nutrition session had the unifying theme of meat quality, animal health and production efficiency. Consumers judge the quality of meat by its appearance, texture and flavour. The most important factor at the point of sale is appearance and this can be greatly affected by lipid oxidation. With increasing consumer demand for unsaturated fatty acids in meat, lipid oxidation becomes more of a problem. Evidence presented in the first paper shows that supplementation of diets with vitamin E can significantly improve the keeping qualities of meat in poultry, pigs and cattle. Another consumer concern is the presence of antibiotics in meat. Feed antibiotics for pigs have been banned in Sweden and the third paper looks at how attention to detail can maximise the efficiency of production without growth promoters. Of particular importance are correctly balanced rations, health status, housing and management. The other paper in this session concentrated on the health of young animals; in particular, the response of the immune system to plant proteins. It is postulated that systemic hypersensitivity and allergic reactions can be reduced or eliminated by strategic exposure to lectins.

The Poultry session comprised two papers on topical subjects. The first of these addressed the problem of wheat interfering with the digestion of fats. High inclusion levels of cereals can increase digesta viscosity and inhibit the movement of bile salts and lipase within the intestines. This problem can be overcome to some extent by including enzymes in the diet to reduce viscosity. In turkeys, continued genetic selection for growth rate has led to birds reaching market weights at increasingly younger ages. Since breast meat is the most valuable part of the carcass, it is important to look at slaughter ages which will give optimum yields of this portion. The relationships between growth rate, slaughter age and breast-meat yield were examined in the second paper of this session.

The Ruminant session was entirely devoted to dairy cows this year. Dairy farmers in most parts of the developed world are currently facing reductions in milk prices with potentially increasing input costs. It is therefore important that we continue to examine the efficiency of utilisation of raw materials and production systems in general. The first paper examined the role of fats in dairy cow diets. Fats are high-energy sources that have a vital role to play in providing sufficient energy for the modern dairy cow. However, it is important to appreciate the effects they may have on rumen fermentation and production responses. The potential for modifying the fat composition of milk through manipulation of dietary fats is also explored. This could be one way of increasing product value by supplying milk for niche markets. The second paper discussed the use of starch in dairy rations. Like fats, starch can increase the energy content of diets, but it can also have negative effects on forage digestibility. The potential for directly increasing glucose supply by feeding starch that bypasses rumen fermentation is discussed, as is the effects of starch on milk protein concentration. Other papers in this session relate developments in the French INRA system of feeding dairy cows and amino acid nutrition. These two papers are complementary and both provide a great deal of useful information that should improve the accuracy of meeting the requirements of high-yielding cows. The INRA system has been developed into a more integrated system that now includes net energy, metabolisable protein and intake prediction. The paper on amino acids shows how ration formulation for dairy cows is approaching the level precision enjoyed by non-ruminant nutritionists. The other paper in this session analysed systems of feeding and managing high-yielding dairy cows in the US. Cows in the US are characterised by high genetic merit and to reach their potential it is necessary to pay careful attention to nutrition and feed ingredients. Added boosts to yield come from the increasing use of frequent milking and BST. These provide even greater challenges for feed formulators and the shortcomings in current feed evaluation systems are discussed.

The overall themes of this conference were product quality and improving efficiency through accurate knowledge of the nutritional value of raw materials. All of the papers are topical and could have immediate application in the animal feed industry.

We would like to thank the speakers for their oral and written contributions. We would also like to thank Trouw Nutrition for their continued financial support of the conference. Finally, we would like to acknowledge the contribution of Dr Will Haresign to the organisation of the conference. He acted as Secretary from 1976 until 1991, served on the Programme Committee until he took over from Dr Des Cole as Chairman in 1996 and finally left Nottingham after the 1997 conference to take up the Chair of Agriculture at Aberystwyth. The animal feed industry is extremely grateful to him.

P.C. Garnsworthy
J. Wiseman

CONTENTS

I

General Nutrition

1

EFFECTS OF VITAMINS IN THE FEED ON MEAT QUALITY IN FARM ANIMALS: VITAMIN E

P.J.A. SHEEHY[1], P.A. MORRISSEY[1], D.J. BUCKLEY[2] and J. WEN[3]

Departments of [1]Nutrition, and [2]Food Technology, University College, Cork, Ireland, and [3]Institute of Animal Science, Chinese Academy of Agricultural Science, Malianwa, Haidan, Beijing, China.

Introduction

The three sensory properties by which customers judge meat quality are appearance, texture, and flavour (Liu *et al.*, 1995). The most important of these properties is the appearance because this strongly influences the initial decision of the customer to purchase or reject the product. One of the main factors which limits the acceptability of meat and meat products is the process of lipid oxidation. In raw meat, this results in the formation of brown pigments (especially in beef), increased drip losses and the development of unacceptable odours, while in cooked and stored meats it causes off-flavours, such as 'warmed-over' flavour. As well as having a negative impact on the acceptability of meats and meat products, lipid oxidation may also have food safety implications. Concerns have been expressed over the possible atherogenic effects of lipid oxidation products (e.g. malondialdehyde and cholesterol oxides) in the body (Addis and Warner, 1991). Preventing lipid oxidation during processing, storage and retail display is therefore essential in order to maintain the quality, wholesomeness and safety of meats, and to ensure that customers will make repeat purchases. However, the oxidative stability of meat is low and difficult for retailers to predict accurately (especially in the case of poultry or fish, which are rich in polyunsaturated fatty acids - PUFA), and oxidation often results in meat being discounted or even removed from retail display before the end of its normal bacteriological shelf-life.

Lipid oxidation

Lipid oxidation is a process in which PUFA are attacked by highly-reactive free radicals and give up a loosely-bound hydrogen atom from an allylic CH_2 group

(Kappus, 1991), leaving behind a fatty acid radical. This reaction can also be brought about by irradiation. Fatty acid radicals can have several fates, the most likely of which in aerobic cells is molecular rearrangement to a conjugated diene, followed by reaction with O_2 to produce a peroxyl radical. If this peroxyl radical attacks another polyunsaturated fatty acid, a hydroperoxide and a second fatty acid radical are formed, and the latter can then begin the process over again, setting up a chain-reaction which could continue until all of the PUFA are used up. Hydroperoxides can be broken down by heating, irradiation or under catalysis by metal ions (e.g. Fe^{++} or Cu^+) to give a range of compounds such as aldehydes, alkanes, acids and alcohols, many of which possess objectionable flavours and odours. A typical example is the 'warmed-over' flavour which is found in cooked meats after refrigerated storage and reheating (Asghar *et al.*, 1988).

It is generally believed that lipid oxidation in meats is initiated in the phospholipid fraction in subcellular membranes. This fraction is characterised by the presence of relatively high levels of PUFA. The process probably begins immediately after slaughter, since the biochemical changes in the muscle that accompany post-slaughter metabolism and ageing give rise to conditions where oxidation is no longer tightly controlled and the balance of prooxidative factors/antioxidative capacity favours oxidation (Morrissey *et al.*, 1994). There is still confusion as to the nature of the initiation process. Spontaneous fatty acid radical formation or direct reaction of unsaturated fatty acids with molecular oxygen are thermodynamically unfavourable (Hsieh and Kinsella, 1989). Instead, it appears that transition metals, especially iron, are pivotal in facilitating the generation of radical species capable of abstracting loosely-bound hydrogens from the PUFA (Gutteridge and Halliwell, 1990a,b; Kappus, 1991; Kanner *et al.*, 1992). A small "transit pool" of iron exists in tissues chelated to nucleotides, amino acids, and inorganic and organic phosphates (Dunford, 1987). These low molecular weight water-soluble chelates are thought to be responsible for the catalysis of lipid oxidation in biological tissues (Halliwell and Gutteridge, 1986). Iron also exists in the form of high molecular weight fractions associated with membrane lipids and proteins such as haemoglobin, myoglobin, ferritin and haemosiderin. Some high molecular weight iron sources such as haemoglobin and myoglobin can directly catalyze lipid oxidation (Tichivangana and Morrissey, 1985; Apte and Morrissey, 1987; Harel and Kanner, 1985; Decker *et al.*, 1993; Monahan *et al.*, 1993a), while others can do so once their iron is released into the low molecular weight pool (Decker *et al.*, 1993). Thus, iron released from ferritin during storage, processing and cooking may also catalyze lipid peroxidation in muscle foods (Apte and Morrissey, 1987; Decker and Welch, 1990; Kanner and Doll, 1991; Decker *et al.*, 1993). The relative contributions of the different forms of iron in catalysing lipid peroxidation have not been clearly defined. The rate and extent of oxidation may also be influenced by pre-and post-slaughter events such as stress, rate of pH decline, or carcass temperature, and by disruption of muscle membrane integrity by mechanical deboning, mincing, restructuring or cooking.

Lipid oxidation can be prevented or at least retarded in several different ways. In living animals, an elaborate enzymatic system (consisting of glutathione peroxidase, glutathione reductase, catalase and superoxide dismutase) protects against the formation of free radical species which could initiate the oxidation reaction (Halliwell, 1987; Machlin and Bendich, 1987). However, the cessation of blood flow and oxygen supply to the muscle after slaughter results in the accumulation of lactic acid (the end-product of post-mortem glycolysis) in the tissue. This lowers the pH from near neutrality to approximately pH 5.5, and it is unlikely that the antioxidant enzyme system still functions under these acidic conditions (Morrissey *et al.*, 1994). Exclusion of oxygen by inert gas or vacuum-packing is an obvious means of preventing lipid oxidation in meats and processed meat products. However, this avenue is only feasible to a limited extent, and residual oxygen levels of less than 1% are extremely difficult to obtain in a production environment (Löliger, 1991). Stabilization of processed meat products usually requires additional measures, such as the use of nitrites (Morrissey and Tichivangana, 1985), metal-chelating agents (e.g. phosphates, EDTA and citrates) (Sato and Hegarty, 1971), and synthetic antioxidants (e.g. BHA, BHT and TBHQ) (Chastain *et al.*, 1982; Crackel *et al.*, 1988). The latter interrupt the free-radical chain reaction by providing an alternative source of hydrogens.

In recent years, resistance to the use of synthetic antioxidants in foods has increased, and there is now a great deal of interest in replacing them with natural substances, many of which possess antioxidant properties. Numerous authors have clearly demonstrated that dietary supplementation with vitamin E increases the stability of lipids in meat systems against off-flavour development, myoglobin oxidation, and other manifestations of lipid breakdown. There is also interest in the potential antioxidant effects of other substances such as ascorbic acid, β-carotene, glutathione, carnosine, homocarnosine and anserine.

Vitamin E in animal nutrition

Compounds with vitamin E activity are called tocopherols and tocotrienols. There are four forms of each, the most important from a biological viewpoint being α-tocopherol. However, this form is rather unsuitable for incorporation into vitamin premixes because of its oily consistency and poor stability, and esters, such as α-tocopheryl acetate, are most commonly used. Indeed, because of its widespread use in vitamin premixes, the International Unit (IU) of vitamin E activity is based on synthetic *all-rac*-α-tocopheryl acetate (an equimolar mixture of all 8 possible stereoisomers), rather than the naturally-occurring RRR-α-tocopherol. One IU of vitamin E is the activity of 1 mg of *all-rac-α*-tocopheryl acetate, 0.735 mg RRR-α-tocopheryl acetate, 0.671 mg RRR-α-tocopherol, or 0.909 mg *all-rac-α*-tocopherol (NRC, 1994). Following ingestion, the ester bond of α-tocopheryl acetate is hydrolysed by a specific esterase in the gut (Mueller *et al.*, 1976), releasing the free α-tocopherol for absorption and transport to the liver in chylomicrons *via* the lymphatic system (Bjornebøe *et al.*, 1986), or, in the case of birds, *via* the hepatic portal vein (Gallo-Torres, 1980). It is then exported in lipoproteins to the muscles and other tissues (Bjornebøe *et al.*, 1987) where it appears to be laid down in specific locations in membranes, in close proximity to membrane-bound enzyme systems and their associated PUFA (Molenaar *et al.*, 1980). In living animals, vitamin E functions as a chain-breaking antioxidant in the cellular and subcellular membranes, where it quenches free radicals arising during normal metabolism (McKay and King, 1980).

The absence of vitamin E from the diet results in several characteristic deficiency symptoms, including mulberry-heart disease and liver necrosis in pigs, white-muscle disease in ruminants, and exudative diathesis, nutritional muscular dystrophy, and nutritional encephalomalacia in poultry. The recommended dietary vitamin E concentrations given in feeding tables may simply be the minimum levels needed to prevent these conditions (called the vitamin E 'requirement') or may also incorporate a safety factor, in which case it is referred to as the 'vitamin E allowance'. Requirements and allowances vary between species, and between different ages of animal within the same species. Vitamin E requirements may also be influenced by the dietary concentration of selenium (required for optimal activity of gluthathione peroxidase), the concentration and type of fat in the feed, and the presence of pro- or antioxidants. Examples of the vitamin E requirement of different species of meat animals and fish are given in Table 1.1. In general, the values are small, with the obvious exceptions being those for fish.

Tables of nutrient requirements are sometimes of limited value for feed formulators since the studies upon which they are based are carefully controlled and animals may not have encountered the usual stresses (crowding, transport, etc.) to which they would be exposed in the commercial environment (Ward, 1993). Furthermore, body weight gains and productive yields have increased substantially in recent years, whereas some of the studies upon which requirements are based may date from the 1950's or 1960's. Consequently, it is not unusual for feed formulators to exceed published requirements substantially as a safety factor.

McIlroy *et al.* (1993) reported that the average vitamin E concentration in feeds given to 168 broiler flocks in Northern Ireland was 48 IU/kg diet, compared with a requirement of 10 IU/kg (NRC, 1994). However, Ward (1993) surveyed more than 90% of feed producers in the United States and reported that vitamin E levels in broiler and turkey feeds there were considerably lower. Average vitamin E concentrations in broiler starter, grower and finisher feeds were 16.3, 14.3 and

12.4 mg/kg, but the mean of the lowest 25% of values in the grower and finisher feeds were only 8.60 and 6.27 mg/kg. Similarly, the average vitamin E concentrations in turkey starter, grower and finisher feeds were 34.7, 24.5 and 14.4 mg/kg, compared with a requirement of approximately 10-12 IU/kg, but the means of the lowest 25% of values in the grower and finisher feeds were only 9.97 and 8.14 mg/kg, respectively. Although vitamin E was being added by hand in some locations for meat preservation purposes, it was clear that levels in many broiler and turkey feeds produced in the US at that time were quite low. Indeed, the levels in poultry feeds in Northern Ireland are also several-fold lower than those now known to be most effective in stabilizing meat lipids. The situation could be the same in commercial pig, cattle and fish feeds, although precise data are lacking.

Table 1.1 VITAMIN E REQUIREMENTS OF SOME MEAT ANIMALS AND FISH (IU/KG)

Species	*Vitamin E requirements (IU/kg diet)*
Beef Cattle[a]	
Calves	15-60
Adult	Normal diet apparently adequate
Pigs[b]	
Starting (1-10 kg)	16
Growing (10-50 kg)	11
Finishing (50-110 kg)	11
Poultry[c]	
Broiler chickens	
0-3 Weeks	10
3-6 Weeks	10
6-8 Weeks	10
Turkeys	
0-8 Weeks	12
8-20/24 Weeks	10
Ducks	
0-2 Weeks	10
2-7 Weeks	10
Fish[d]	
Atlantic salmon	39
Rainbow trout	27.5-110
Channel catfish	27.5-55
Common carp	110

[a] NRC, 1984, [b] 1988, [c] 1994, [d] 1993.

Dietary vitamin E supplementation and meat quality

CHICKENS

One of the earliest reports showing the ability of vitamin E to act as an antioxidant in meats was by Marusich *et al.* (1975). These authors fed a diet containing 16 IU α-tocopherol/kg to a control group of broilers for 8 weeks, while other groups were given additional supplements of either 4, 14, 24 or 44 IU *all-rac* α-tocopherol/ kg. Concentrations of α-tocopherol in breast muscle increased from 2.0 to 6.2 mg/kg as dietary α-tocopherol increased and there was a corresponding decrease in the extent of lipid oxidation in breast muscle after storage of the whole carcasses at 1°C. The TBA number (an indicator of lipid oxidation) of breast muscle from the control group was 0.86 after 5 days whereas this value was only 0.36 after 14 days for the group fed the highest level of vitamin E. Studies from this laboratory provided further evidence of a protective effect of vitamin E. In chicks given a basal diet (5 mg α-tocopherol/kg) or the same diet supplemented with α-tocopheryl acetate (final α-tocopherol concentration 25, 65 and 180 mg/kg) for 32 days, the α-tocopherol concentration in thigh muscle increased directly in proportion to the concentration of α-tocopherol in the diet (Sheehy *et al.*, 1991), and the susceptibility of thigh muscle to iron-ascorbate induced lipid oxidation was significantly reduced (Sheehy *et al.*, 1990). Supplementation also increased the oxidative stability of both raw and cooked thigh muscle during frozen storage (Sheehy *et al.*, 1993a).

Short-term feeding of broilers with 160 IU α-tocopherol/kg for the last 5 days prior to slaughter was effective in retarding the onset of rancidity in raw whole breast muscle (Marusich *et al.*, 1975), suggesting that a high vitamin E supplement could be given in the finishing diet as an alternative to continuous supplementation. However, considering the relatively slow uptake of α-tocopherol into chicken muscle compared with other tissues (Sheehy *et al.*, 1991) and the apparent requirement to lay down the vitamin in specific locations within the muscle membranes for optimum protection, some authors have questioned whether short-term supplementation would guarantee adequate stability in processed muscle. Bartov and Frigg (1992) fed broilers a basal diet containing no added vitamin E for 7 weeks (treatment 1), the basal diet supplemented with 100 mg vitamin E/kg for 7 weeks (treatment 2), basal diet supplemented with 150 mg/kg to 3 weeks and basal diet alone to 7 weeks (treatment 3), basal diet with 150 mg/kg, 0 and 100 mg/kg to week 3, from weeks 3 to 6 and from 6 to 7 weeks (treatment 4) and the basal diet alone to 5 weeks and then basal diet plus 100 mg/kg to 7 weeks (treatment 5). The stability of meat from treatments 3, 4 and 5 was significantly greater than that from treatment 1 (no vitamin E) but it was significantly lower than that from treatment 2 (vitamin E given continuously). Brandon *et al.* (1993) gave one group of broilers a basal diet containing 30 mg α-tocopheryl acetate/kg feed continuously up to slaughter at 6 weeks, while other groups were given a supplemented diet containing 200 mg α-tocopheryl acetate/kg for 1, 2, 3, 4 or 5 weeks immediately prior to slaughter. The α-tocopherol content of leg and breast muscle increased as the pre-slaughter supplementation period increased from 0 to 5 weeks (Figure 1.1). Supplementation of broiler diets with 200 mg α-tocopheryl acetate/kg for 5 weeks prior to slaughter improved the oxidative stability of ground muscle during refrigerated and frozen storage and protected against the pro-oxidant effect of salt (Figure 1.2). These results suggest that a pre-slaughter supplementation period of at least 4-5 weeks, feeding 200 mg α-tocopheryl acetate/kg (i.e. some 20 times higher than the NRC requirement), is necessary to attain the optimum protective benefit of α-tocopherol in processed meat. Since oxidative stability is also influenced by other factors, such as the amount (NRC, 1994) and type (Lin *et al.*,1989a; Asghar *et al.*, 1990; Huang and Miller, 1993) of fat in the diet, as well as the fat quality (Lin *et al.*, 1989b; Sheehy *et al.*, 1993b,c), these figures may need to be adapted in certain situations.

Lipid oxidation affects not only the PUFA in muscle membranes, but also the cholesterol, and concern about the presence of cholesterol oxidation products (COPs) in foods is increasing because of the possibility that they may be involved in the primary events leading to atherosclerosis (Steinberg *et al.*, 1989). Production of COPs in chicken meat varies greatly according to the cooking method and the type of portion (Chen *et al.*, 1993). Breast and thigh muscle samples which were boiled for 4 h contained 20-hydroxycholesterol, cholesterol α-epoxide and cholestane triol, while deep fat frying and microwave cooking produced 20-hydroxycholesterol. Roasting did not result in the production of COPs, except in the skin. Galvin *et al.* (1995) gave broilers diets supplemented with 20 (control), 200 and 800 mg α-tocopheryl acetate/kg for 6 weeks and measured COPs production in patties manufactured from ground breast or thigh muscle immediately after cooking and, additionally, following 6 and 12 days refrigerated storage. 25-Hydroxycholesterol was present at all times, 20 α-hydroxycholesterol was detected in breast muscle-based patties on day 6, and 7-ketocholesterol was detected in samples from breast and thigh on day 12. In general, COPs production was slow. However, after 12 days storage, a significant protective effect of vitamin E supplementation was observed. Total COPs concentrations in patties from broilers fed 200 mg α-tocopheryl acetate/kg were about 50% of those in control samples, while supplementation with 800 mg/kg reduced total COPs to about 33% of control values.

Lipid hydroperoxides break down to aldehydes, alkanes, and other compounds, many of which possess unacceptable flavours. Dietary vitamin E supplementation should, therefore, be capable of reducing the appearance of off-flavours in meats

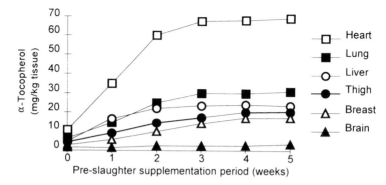

Figure 1.1 α-Tocopherol concentrations in heart, liver, lung, brain, thigh and breast muscle of chicks fed a basal diet of 30 mg α-tocopheryl acetate/kg up to slaughter at 6 weeks, or chicks fed an α-tocopheryl acetate-supplemented diet (200 mg/kg) for 1-5 weeks immediately prior to slaughter. Values are means of 8 samples analysed in duplicate (Courtesy of Brandon *et al.*, 1993).

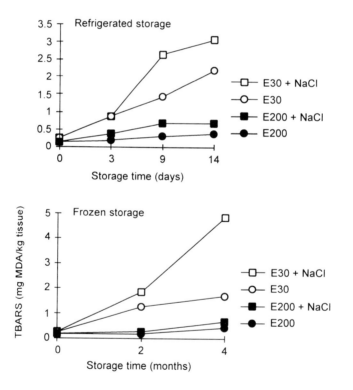

Figure 1.2 Effect of dietary α-tocopheryl acetate supplementation on lipid oxidation in ground broiler leg muscle during refrigerated and frozen storage. Broilers were fed a basal diet containing 30 mg α-tocopheryl acetate/kg for 6 weeks or an α-tocopheryl-acetate supplemented diet (200 mg/kg) for the last 5 weeks prior to slaughter. The oxidative stability was determined in the presence or absence of 1% salt (Courtesy of Brandon *et al.*, 1993).

and meat products. Up to now, researchers have tended to focus more on measuring chemical indices of lipid oxidation (TBARS, etc.) than on sensory properties. However, Ristic (1991) made use of sensory evaluation following a 35-day feeding trial in which chicks were fed diets containing 0, 40 or 80 mg α-tocopheryl acetate/ kg continuously, or 40 mg/kg increasing to 150 mg/kg for the final week before slaughter. The carcasses were stored at 0° C for 1, 5 or 12 days. Juiciness of the breast and thigh meat was best with the 40/150 mg/kg vitamin E supplement. Tenderness, aroma and general acceptability of the breast were best for the 40 and 80 mg/kg (continuous) treatments. Tenderness of the thigh was best for the 80 mg/kg and 40/150 mg/kg treatments. Blum *et al.* (1992) reported that the flavour of meat from broilers fed a control diet containing 20 mg α-tocopherol/kg feed deteriorated significantly during 12 days storage at 4°C, whereas flavour remained unchanged in meat from birds fed a supplemented diet (160 mg/kg) up to the end of the storage period. In contrast, no significant effect of dietary vitamin E supplementation on sensory properties of meat samples stored at -18°C for 18 months was observed. O'Neill *et al.* (1995) studied the effect of vitamin E supplementation of tallow- or olive oil-based broiler diets on the development of warmed-over flavour in cooked thigh muscle. Supplementation with 200 mg α-tocopheryl acetate/kg for 8 weeks significantly increased oxidative stability and delayed the onset of warmed-over flavour in cooked ground muscle during refrigerated storage for up to 5 days. There was also a significant protective effect against both lipid oxidation and warmed-over flavour when frozen meat was thawed, cooked, and stored for 2 days in the refrigerator.

TURKEYS

The efficacy of dietary vitamin E supplementation as a natural protective mechanism against oxidation has been demonstrated in whole turkey muscle (Marusich *et al.*, 1975; Sklan *et al.*, 1983), turkey meat composites (Sheldon *et al.*, 1984), and carcass fat (Franchini *et al.*, 1990). However, α-tocopherol concentrations in turkey tissues are considerably lower than those found, for example, in broilers (Marusich *et al.*, 1975; Sklan *et al.*, 1982), and this must be taken into account in establishing effective dietary vitamin E levels for turkey feeding.

Wen *et al.* (accepted for publication) examined the uptake and distribution of α-tocopherol in turkeys fed either a basal diet supplemented with 20 mg α-tocopheryl acetate/kg or high vitamin E diets containing 300 or 600 mg α-tocopheryl acetate/kg, and reported that there were marked differences in the concentrations of α-tocopherol in different tissues: gizzard contained the highest concentrations, while low levels were present in muscle, fat and brain. The time-course of α-tocopherol uptake by the breast and thigh muscle is shown in Figure 1.3. In controls, α-tocopherol concentrations fell sharply between weeks 1 and 3 and remained low for the duration of the experiment. This finding is of interest in view of the fact that the control diet was supplemented with 20 mg α-tocopheryl acetate/kg, approximately twice the NRC requirement. Supplementing the diet with 300 mg α-tocopheryl acetate/kg permitted α-tocopherol to accumulate slowly in muscle up to week 13, after which the levels stabilized, while feeding 600 mg/kg allowed α-tocopherol concentrations to increase at a faster rate up to week 13, after which they, too, stabilized. Thus, it takes approximately 13 weeks for α-tocopherol concentrations in turkey thigh and breast muscle to reach a plateau in response to high dietary intakes.

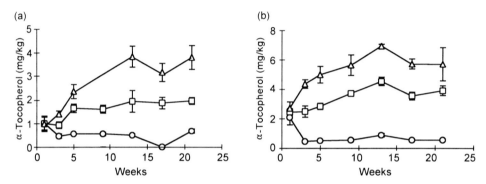

Figure 1.3 Time-course of α-tocopherol uptake in (a) breast muscle and (b) thigh muscle of turkeys fed diets supplemented with 20, 300, or 600 mg α-tocopheryl acetate/kg. Values are means and standard deviations from 3 determinations (Courtesy of Wen *et al.*, accepted for publication).

Vitamin E at dietary levels of 300 and 600 mg/kg was clearly protective against iron-ascorbate induced oxidation (Wen *et al.*, accepted for publication). Regression analysis indicated that increasing breast muscle α-tocopherol to about 2.0-5.0 mg/kg and thigh muscle α-tocopherol to 4.0-8.0 mg/kg reduced the susceptibility of the muscle to induced lipid oxidation. α-Tocopherol concentrations in raw burgers from turkeys fed 300- and 600 mg α-tocopheryl acetate/kg were 6.1- and 9.8-fold higher, respectively, compared with those from the control turkeys (Figure 1.4a) (Wen *et al.*, 1996). The concentrations of α-tocopherol observed in the latter were about one-fifth those made from muscle of broilers fed a broadly similar basal diet supplemented with 30 mg α-tocopheryl acetate/kg (Brandon *et al.*, 1993). Cooking (70°C X 60 min) did not significantly affect the α-tocopherol content of the burgers, but TBA numbers in the cooked burgers were roughly twice those in raw burgers (Figure 1.4b). TBA numbers of both raw and cooked burgers from turkeys fed the control diet were significantly higher than those from turkeys fed 300- or 600 mg α-tocopheryl acetate/kg. During refrigerated storage over a 6-day period, lipid oxidation was more pronounced in burgers from turkeys given the control diet. Values for the supplemented groups were less than half of those of the controls. Similar results were observed during frozen storage over 6 months.

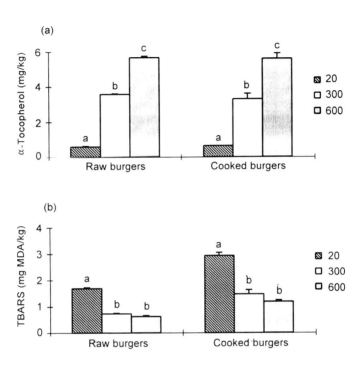

Figure 1.4 Effect of dietary supplementation with 20, 300, or 600 mg α-tocopheryl acetate/kg for 21 weeks on concentrations (mg/kg) of (a) α-tocopherol and (b) thiobarbituric acid-reactive substances (TBARS, expressed as malondialdehyde, MDA) in raw and cooked burgers made from turkey breast muscle. Values are means and standard errors from 3 determinations. For each product and parameter, means not sharing a common superscript letter are significantly different (P < 0.01) (After Wen *et al.*, 1996).

CATTLE

A number of recent studies in the United States have provided evidence that vitamin E supplementation of cattle diets delays lipid oxidation in fresh meat during display under simulated retail conditions. Ground, frozen and cooked beef can also be protected against lipid oxidation by dietary vitamin E supplementation. Furthermore, it quickly became apparent during studies that the improvement in lipid stability was associated with a marked enhancement of colour (myoglobin) stability. The precise mechanisms linking vitamin E with the inhibition of myoglobin oxidation have not been fully elucidated, but the colour-stabilizing effect of vitamin E has attracted great commercial interest because of the potential economic benefits.

Fresh beef is cherry-red in colour, but this is followed by a brownish discolouration of the meat surface which may be interpreted as an indication of unwholesomeness (Liu *et al.*, 1995). Beef that becomes discoloured is often minced and sold at a reduced price, and, therefore, factors which can reduce the rate of discolouration are of considerable economic importance to retailers. Beef colour is principally due to the redox state of myoglobin in the muscle fibres, and the nature of the ligand bound to the myoglobin (Liu *et al.*, 1995; Schaefer *et al.*, 1996). Deoxymyoglobin is the purple pigment observed in freshly cut meat. Following several minutes exposure to air, this becomes oxygenated to oxymyoglobin, taking on a bright red colour. After prolonged exposure to air, oxymyoglobin is converted to metmyoglobin, in which water replaces the oxygen. Both deoxymyoglobin and oxymyoglobin are heme proteins in which iron exists in the ferrous (Fe^{++}) form, while metmyoglobin (MetMb) contains ferric (Fe^{+++}) iron. The conversion of ferrous to ferric iron is brought about by oxidation, which may be caused by the direct action of pro-oxidants, or could be mediated by lipid radicals and peroxides (Gutteridge and Halliwell, 1990a,b; Minotti and Aust, 1992; Schaich, 1992).

Several important studies have demonstrated that vitamin E supplementation is very effective in reducing the oxidation of myoglobin in fresh, ground, and frozen beef muscle, and these have been comprehensively reviewed by Liu *et al.* (1995) and Schaefer *et al.* (1996). Faustman *et al.* (1989a) compared the colour stability of sirloin steaks from Holstein steers given a vitamin E-supplemented or unsupplemented high moisture corn/corn silage diet for 10 months. The vitamin E supplement was 370 IU/animal/day. Chroma and Hunter 'a' values of steaks tended to decrease over the course of an 8-day refrigeration period under cool white fluorescent lighting, but values for steaks from the supplemented group were significantly greater than those from the unsupplemented animals, indicating that vitamin E stabilized the redness and colour intensity of the steaks. Arnold *et al.* (1992) fed Holstein steers 300 IU/day of supplemental vitamin E in a high moisture corn-based diet for 9 months before slaughter and reported that the display life of *Longissimus lumborum* steaks was increased by vitamin E supplementation compared with that of steaks from control animals. Steaks from supplemented steers were acceptable to a panel for 7.4 days, compared with 4.9 days for the control steaks. In a subsequent study, calculated colour display life for 7-day-aged *Longissimus lumborum* and *Gluteus medius* was increased by about 50 and 75%, respectively, by dietary supplementation with 360 IU α-tocopheryl acetate/day for 252 days (Arnold *et al.*, 1993a). Supplementation with 1290 IU daily did not give additional benefits for 7-day aged beef but resulted in the most colour-stable meat when an ageing period of 21 days was employed. Colour display life of 7-day aged *Longissimus lumborum* from Holstein steers was extended by 3.4 to 5.7 days following supplementation with 1840 to 3610 IU vitamin E/day for the last 100 days before slaughter (Arnold *et al.*, 1993b), while display life of 7-day aged *Longissimus lumborum* and *Gluteus medius* was increased by 4.8 and 1.6 days, respectively, following supplementation of 1266 IU/day for the last 38 days before slaughter (Arnold *et al.*, 1992).

Ground beef represents about 45% of total beef consumption in the United States (Liu *et al.*, 1995). However, it tends to become brown and rancid more rapidly than whole muscle cuts, because grinding exposes a greater surface to the air and microbial contamination, and accelerates the loss of intracellular reductants such as NADH which help to minimize metmyoglobin formation (Mitsumoto *et al.*, 1991). Faustman *et al.* (1989b) demonstrated that vitamin E supplementation (370 IU/day for approximately 10 months) of Holstein steers significantly increased the colour stability of beef patties made from *Gluteus medius*. Similarly, Mitsumoto *et al.* (1993) showed that vitamin E supplementation (1500 IU/day for 232 days) resulted in lower initial and final percentages of metmyoglobin (from 6.8 to 40.4% MetMb) in Holstein beef patties during 9 days storage compared with control samples (19-86.8% MetMb). Based on a cut-off point of 30%, above which ground beef is usually rejected, these authors estimated that the extension in shelf-life due to vitamin E supplementation was of the order of 6 days. Lavelle *et al.* (1995) reported that the retail display life of ground beef from Holstein steers supplemented with 500 and 2000 IU vitamin E/day for 120 days prior to slaughter was increased by 12 and 42 h, respectively, as judged by a visual panel.

Marketing frozen beef has advantages in terms of cost, convenience and long-term stability, but can sometimes result in deterioration in appearance of the meat (Lanari *et al.*, 1993). These authors investigated the effect of vitamin E supplementation of Holstein steers (2100 IU/day for 126 days) on the colour stability of beef during storage at -20°C, either in the dark or under cool white fluorescent lighting. Supplementation increased the α-tocopherol content of the muscle more than 10-fold, and the colour of the supplemented *Longissimus lumborum* was more stable than than from unsupplemented animals. Considering a saturation index of 16 as representing the limit of acceptability for frozen beef, the colour of control samples was unacceptable after 1 day of dark storage or illuminated display, whereas the colour shelf-life of supplemented beef under these conditions was 214 and 101 days, respectively.

Mitsumoto *et al.* (1993) compared the efficacy of α-tocopherol incorporated into beef muscle by dietary supplementation with that added as a postmortem supplement. Strip loins from Holstein steers that had received 0 or 1500 mg α-tocopheryl acetate/day for 242 days were ground, and the α-tocopherol concentration of the control meat was adjusted by exogenous addition so that it was similar to that from the supplemented group. Endogenous vitamin E improved lipid and pigment stability much more effectively than did exogenous vitamin E, suggesting that the proper positioning of α-tocopherol in the cellular and subcellular membranes rather than on the external surface of the muscle is critical.

Liu *et al.* (1995) suggested that the ideal vitamin E supplementation strategy for cattle would be to feed the vitamin at a level which increases muscle α-tocopherol sufficiently almost to maximize its protective effects. Attempts to saturate the muscle fully with α-tocopherol would be excessively costly for only a small additional benefit. Faustman *et al.* (1989b) suggested that the ideal concentration of α-tocopherol in *Gluteus medius* is 3 mg/kg, while Arnold *et al.* (1993b) proposed that *Longissimus lumborum* α-tocopherol concentration should ideally be about 3.3 mg/kg. Liu *et al.* (1995) recommended that cattle for the domestic market should be supplemented with 500 IU of vitamin E/day for 126 days prior to slaughter in order to attain muscle α-tocopherol concentrations of this order.

PIGS

The average fat content of pig carcasses in the UK has fallen by about 0.5% per year over the last 20 years, mainly due to genetic improvements achieved by pig breeding companies (Warkup, 1994). One of the consequences of this genetic improvement has been that pig fat has become less saturated. The ratio of PUFA to saturated fatty acids is now approximately 0.44, compared with values close to 0.10 and 0.05 in lamb and beef, respectively (Warkup, 1994). This softer fat is more susceptible to oxidative damage, and this may cause difficulties for the major retailers who are increasingly turning towards centralised butchery and modified atmosphere packaging, both of which lead to the meat being exposed to higher levels of oxygen for a longer time prior to retail.

Monahan *et al.* (1993b) observed a significant increase in the α-tocopherol content of *Longissimus dorsi* muscle of pigs fed a diet containing 200 mg α-tocopheryl acetate/kg for 7 days, compared with levels in pigs fed a control (10 mg/kg) diet for the same period. However, uptake of vitamin E by adipose tissue was slower, and it was necessary to feed the supplemented diet for 39 days before the α-tocopherol content of the subcutaneous adipose tissue increased significantly above that of pigs fed the control diet. Vitamin E supplementation reduced lipid oxidation and improved colour stability of pork chops during simulated retail display (Monahan *et al.*, 1992a). TBARS values were lower, and surface redness (Hunter 'a' value) was higher in pork chops from pigs given 100 or 200 mg α-tocopheryl acetate/kg diet, compared with pigs fed 10 mg/kg diet after 2, 4, 6 and 8 days of refrigerated storage. These supplemental levels of vitamin E also reduced the concentrations of β-epoxide, 7β-hydroxycholesterol, 7-ketocholesterol and total COPs in cooked pork after refrigerated storage for 2 days, compared with pork from pigs fed the control diet (Monahan *et al.*, 1992b).

Some authors have noted that supplementation of pig diets with vitamin E is associated with an improvement in certain sensory properties of pork muscle during storage or retail display. Drip losses from thawing pork steaks were lower in muscle from pigs fed on α-tocopheryl acetate-supplemented diets (100 and 200 mg/kg) than in muscle from pigs fed 10 mg/kg (Asghar *et al.*, 1991). Similar results were observed in fresh pork chops by Monahan *et al.* (1994). More recently, Cannon *et al.* (1995) reported that providing supplemental vitamin E (100 mg/kg for 84 days) resulted in significantly lower TBARS and enhanced sensory properties (lower off-flavours, improved palatability and tenderness) in vacuum-packaged, precooked pork during refrigerated storage for periods of up to 56 days, compared with pork from pigs given an unsupplemented diet.

The effects of vitamin E are not confined to the tissue level, but are also apparent at the level of the subcellular membranes, the presumed site of lipid oxidation. Wen *et al.* (submitted for publication) observed that dietary supplementation of 30-35 kg pigs with 200 or 1000 mg α-tocopheryl acetate/kg for 4 weeks resulted in a progressive increase in the α-tocopherol content of whole muscle, mitochondria and microsomes, compared with corresponding values for pigs fed a control (30 mg/kg) diet. Concentrations of α-tocopherol in muscle, mitochondria and microsomes of pigs fed the 1000 mg/kg diet were 3.2-, 6.1- and 5.6-fold greater than those in their counterparts from the control group. The increase in α-tocopherol was associated with a progressive decrease in susceptibility of the whole muscle and subcellular membranes to iron-ascorbate induced lipid oxidation measured by conventional TBARS assay, and this was confirmed by first-derivative spectrophotometry.

Economics of dietary vitamin E supplementation

Generally, vitamin E supplementation has not positively affected performance, carcass characteristics or quality and yield grades of beef cattle (Liu *et al.*, 1995). However, the potential benefits for the beef industry arising out of the ability of vitamin E to stabilize myoglobin could be very considerable. Williams *et al.* (1992) conducted a blind retail study in three locations in the United States. Six different cuts from unsupplemented and vitamin E-supplemented steers (500 IU/day for 100-120 days) were offered for sale over a 4-day period. Across the three locations, losses of original retail value due to discolouration discounts were 5.6 and 2.0% for control and vitamin E-treatments, respectively, a difference of 3.6%. Liu *et al.* (1995) estimated that the cost of supplementing cattle with 500 IU/day of vitamin E for 126 days would be approximately $3 per animal, so the total cost for 25.5 million cattle slaughtered each year in the US would be $76.5 million. Total receipts from fresh beef sales in the US in 1991 were approximately $22 billion.

Extrapolating the improvement in receipts of 3.6%, these authors suggested that the potential gain to the beef industry could be in the order of $792 million, or a benefit:cost ratio of 10.4:1. If cattle feeders were fully compensated and paid an incentive of $3 per animal to feed vitamin E, the benefit:cost ratio would be 5.2:1.

A cost:benefit analysis of this kind to assess the potential benefits of vitamin E supplementation in stabilizing broiler lipids has not yet been carried out. However, the possible costs and benefits in terms of broiler performance were the subject of a study by Kennedy *et al.* (1991). Over 3 million birds from a total of 168 flocks were fed throughout their lives either on normal diets or diets supplemented with vitamin E to contain a total concentration of 180 IU/kg feed. In flocks receiving the supplemented diets, the feed conversion ratio was improved by 0.8% and the average weight per bird was increased by 1.4%, compared with those receiving the normal diets. When the cost of the additional vitamin E was taken into account, the net income per 1000 birds was increased by 2.82%. Although this increase in income was not statistically significant, and therefore could have arisen by chance, the authors suggested that it was economically important on account of the large numbers of birds involved in the study. At the very least, it may be concluded that the cost of vitamin E supplementation was fully recovered, so the beneficial effects on meat quality are likely to be attainable without a significant increase in feeding costs. Subsequently, McIlroy *et al.* (1993) noted that subclinical infectious bursal disease (IBD) was present in approximately half of the flocks, and that analysis of the performance data showed that the average net income of flocks with subclinical IBD and fed a high vitamin E-containing diet was 10% better than that from flocks with subclinical IBD and fed a normal vitamin E-containing diet. In contrast, the difference between the average net income achieved by flocks without subclinical IBD and being fed on either a high or a normal vitamin E-containing diet was only 2% and not significantly different. These authors suggested that the improved performance from high vitamin E-containing diets recorded in flocks with subclinical IBD is due to enhanced immunocompetence and increased resistance to disease. It was also suggested that under field conditions, high dietary vitamin E may be most beneficial where there is a challenge to the defence system of the host and that significantly improved performance would occur more predictably under such conditions.

The lower concentrations of α-tocopherol in turkey muscle almost certainly mean that greater vitamin E inputs and considerably longer supplementation times are necessary to achieve equivalent protection against rancidity, compared with broilers. The extent to which the increased cost of vitamin E inclusion can be justified will ultimately depend on whether or not consistent, statistically significant consumer preferences for high-vitamin E turkey meat and meat products can be demonstrated, and at what dietary vitamin E levels. Relatively little is known concerning the influence of vitamin E supplementation on the sensory properties of turkey meat. The studies of Wen *et al.* (1996) indicate that a satisfactory degree of lipid stability (in terms of reduced levels of TBARS) can be obtained in turkey products by feeding turkeys diets containing 300-600 mg α-tocopheryl acetate/kg continuously. However, before firm recommendations on optimum dietary levels can be made, further studies of this type must be carried out in which the chemical measurements of lipid oxidation are related to sensory evaluation using trained sensory panels and untrained consumers.

Implications for human vitamin E intake

Vitamin E is well tolerated even in pharmacological doses, and relatively few side-effects have been observed in humans even at very high intakes of up to about 3000 IU/day. However, it is important to consider the effect of increasing the vitamin E content of animal diets on likely intakes of the vitamin by consumers. Dietary supplementation with 200 mg α-tocopheryl acetate/kg increases the α-tocopherol concentration of chicken muscle to about 17-20 mg/kg (Brandon *et al.*, 1993), so a typical serving of 80 g of chicken would probably provide about 14-20% of the adult RDA of 8-10 mg/day (NRC, 1989) (assuming no cooking losses). Similarly, burgers made from breast muscle of turkeys fed a high vitamin E diet (600 mg α-tocopheryl acetate/kg for 21 weeks) contained about 5.7 mg α-tocopherol/kg (Wen *et al.*, 1996), and consumption of 1 burger (approximately 90 g) would, therefore, correspond to an intake of about 0.55 mg α-tocopherol, or approximately 5-6% of the RDA. Liu *et al.* (1995) estimated that an 85 g serving of beef containing 3.5 mg α-tocopherol/kg would contribute just 2% of the RDA. Pork from pigs supplemented with 200 mg α-tocopheryl acetate/kg for 67 days contained approximately 6 mg α-tocopherol/kg (Monaghan *et al.*, 1993b), and a serving would, therefore, contribute about 5-6% of the RDA. These figures demonstrate that supplementation of animal diets with vitamin E merely converts meat from being a poor source of the vitamin to being, at best, a moderate one. There do not appear to be any grounds for concern, therefore, about the likely intakes of α-tocopherol in humans consuming vitamin E-enriched meats being excessive.

Conclusion

The vitamin E requirements of animals are influenced by many factors, including species, age and physiological status, the presence or absence of disease, the amount and type of dietary fat, and the interaction of other antioxidants and pro-oxidants.

Meeting these requirements helps prevent deficiency symptoms and maintain overall health. However, judging the nutritional adequacy of animal diets with respect to vitamin E by these criteria alone can have the effect of underestimating by a considerable margin the amount required to ensure optimum meat stability after slaughter, because slaughtering itself, and the many processes to which the meat may then be subjected (mincing, addition of salt, refrigeration, freezing and cooking), provide a considerable amount of oxidative stress. There is now considerable evidence indicating that supplemental vitamin E above physiological levels is very effective in minimizing lipid oxidation, myoglobin oxidation, and, in certain situations, drip losses in meats, which are all indicators of a loss of quality. For broilers, the present evidence suggests that optimal dietary concentrations of vitamin E are in the region of 100 mg (Bruni, 1993) to 200 mg α-tocopheryl acetate/kg (Brandon *et al.* 1993). For turkeys, dietary concentrations of 300 mg α-tocopheryl acetate/kg or more may be necessary for optimum preservation of meat quality because of the slow uptake and lower concentrations achieved in the muscles (Wen *et al.*, accepted for publication). However, this figure needs to be underpinned by further studies, including sensory evaluation. For beef cattle, Liu *et al.* (1995) recommended that they be supplemented with 500 IU (i.e. 500 mg α-tocopheryl acetate)/day for 126 days prior to slaughter, while for pig production, Warkup (1994) stated that the Meat and Livestock Commission is actively encouraging the industry in the UK to include vitamin E at a minimum level of 100 IU/kg diet. These supplemental levels do not expose consumers to high intakes.

Acknowledgement

The authors wish to acknowledge the financial support of F. Hoffmann-La Roche, Basel, Switzerland.

References

Addis, P.B. and Warner, G.J. (1991) The potential health aspects of lipid oxidation products in foods. In *Free Radicals and Food Additives*, pp. 77–119. Edited by O.I. Arouma and B. Halliwell. London: Taylor and Francis.

Apte, S. and Morrissey, P.A. (1987) Effect of water-soluble haem and non-haem iron complexes on lipid oxidation of heated muscle systems. *Food Chemistry*, **26**, 213–222.

Arnold, R.N., Scheller, K.K., Arp, S.C., Williams, S.N., Beuge, D.R. and Schaefer, D.M. (1992) Effect of long- or short-term feeding of α-tocopheryl acetate to Holstein and crossbred beef steers on performance, carcass characteristics, and beef colour stability. *Journal of Animal Science*, **70**, 3055–3065.

Arnold, R.N., Scheller, K.K., Arp, S.C., Williams, S.N. and Schaefer, D.M. (1993a) Dietary α-tocopheryl acetate enhances beef quality in Holstein and beef breed steers. *Journal of Food Science*, **58**, 28–33.

Arnold, R.N., Arp, S.C., Scheller, K.K., Williams, S.N. and Schaefer, D.M. (1993b) Tissue equilibration and subcellular distribution of vitamin E relative to myoglobin and lipid oxidation in displayed beef. *Journal of Animal Science*, **71**, 105–118.

Asghar, A., Gray, J.I., Buckley, D.J., Pearson, A.M. and Booren, A.M. (1988) Perspectives in warmed-over flavour. *Food Technology*, **42**, 102–108.

Asghar, A., Lin, C.F., Gray, J.I., Buckley, D.J., Booren, A.M. and Flegal, C.J. (1990) Effects of dietary oils and α-tocopherol supplementation on membranal lipid oxidation in broiler meat. *Journal of Food Science*, **55**, 46–50.

Asghar, A., Gray, J.I., Booren, A.M., Gomaa, E.A., Abouzied, M.M. and Miller, E.R. (1991) Effects of supranutritional dietary vitamin E levels on subcellular deposition of α-tocopherol in the muscle and on pork quality. *Journal of the Science of Food and Agriculture*, **57**, 31–41.

Bartov, I. and Frigg, M. (1992) Effect of high concentrations of dietary vitamin E during various age periods on performance, plasma vitamin E and meat stability of broiler chicks at 7 weeks of age. *British Poultry Science*, **33**, 393–402.

Bjorneboe, A., Bjorneboe, G.-E. Aa, Bodd, E., Hagen, B.F., Kveseth, N. and Drevon, C.A. (1986) Transport and distribution of α-tocopherol in lymph, serum and liver cells in rats. *Biochimica et Biophysica Acta*, **889**, 310–315.

Bjorneboe, A., Bjorneboe, G.-E. Aa and Drevon, C.A. (1987) Serum half-life, distribution, hepatic uptake and biliary excretion of α-tocopherol in rats. *Biochimica et Biophysica Acta*, **921**, 175–181.

Blum, J.C., Tourraille, C., Salichon, M.R., Richard, F.H. and Frigg, M. (1992) Effect of dietary vitamin E supplies in broilers. 2. Male and female growth rate, viability, immune response, fat content and meat flavour variations during storage. *Archiv fur Geflügelkunde*, **56**, 37–42.

Brandon, S., Morrissey, P.A., Buckley, D.J. and Frigg, M. (1993) Influence of dietary α-tocopheryl acetate on the oxidative stability of chicken tissues. In *Proceedings of the 11th European Symposium on the Quality of Poultry Meat* pp. 397–403. Edited by P. Colin, J. Culioli and F.H. Ricard. Tours: WPSA.

Bruni, M. (1993) Vitamin E and meat quality. *Rivista di Avicoltura*, **62**, 15–19.

Cannon, J.E., Morgan, J.B., Schmidt, G.R., Delmore, R.J., Sofos, J.N., Smith, G.C. and Williams, S.N. (1995) Chemical, shelf-life and sensory properties of vacuum-packaged precooked pork from hogs fed supplemental vitamin E. In *Proceedings of the 41st Annual International Congress of Meat Technology*, pp. 370–371. San Antonio: American Meat Science Association.

Chastain, M.F., Huffman, D.L., Hsieh, W.H. and Cordray, J.C. (1982) Antioxidants in restructured beef/pork steaks. *Journal of Food Science*, **47**, 1779–1782.

Chen, K.H., Yang, S.C. and Su, J.D. (1993) The cholesterol oxidation products contents of chicken meat as affected by different heating methods. In *Proceedings of the 11th European Symposium on the Quality of Poultry Meat*, pp. 412–422. Edited by P. Colin, J. Culioli and F.H. Ricard. Tours: WPSA.

Crackel, R.L., Gray, J.I., Booren, A.M., Pearson, A.M. and Buckley, D.J. (1988) Effects of antioxidants on lipid stability in restructured beef steaks. *Journal of Food Science*, **53**, 656–657.

Decker, E.A. and Welch, B. (1990) Role of ferritin as a lipid oxidation catalyst in muscle foods. *Journal of Agricultural and Food Chemistry*, **38**, 674–677.

Decker, E.A., Crum, A.D., Shantha, N.C. and Morrissey, P.A. (1993) Catalysis of lipid oxidation by iron from an insoluble fraction of beef diaphragm muscle. *Journal of Food Science*, **58**, 233–236, 258.

Dunford H.B. (1987) Free radicals in iron-containing systems. *Free Radicals in Biology and Medicine*, **3**, 405–421.

Faustman, C., Cassens, R.G., Schaefer, D.M., Buege, D.R. and Scheller, K.K. (1989a) Vitamin E supplementation of Holstein steer diets improves sirloin steak colour. *Journal of Food Science*, **54**, 485–486.

Faustman, C., Cassens, R.G., Schaefer, D.M., Beuge, D.R. and Scheller, K.K. (1989b) Improvement of pigment and lipid stability in Holstein steer beef by dietary supplementation with vitamin E. *Journal of Food Science*, **54**, 858–862.

Franchini, A., Giordani, G., Meluzzi, A. and Manfreda, G. (1990) High doses of vitamin E in the turkey diet. *Archiv für Geflügelkunde*, **54**, 6–10.

Gallo-Torres, H.E. (1980) Absorption, transport and metabolism. In *Vitamin E: a comprehensive treatise*, pp. 170–267. Edited by L.J. Machlin. New York: Marcel-Dekker.

Galvin, K., Morrissey, P.A. and Buckley, D.J. (1995) Effect of dietary α-tocopheryl acetate supplementation on the formation of cholesterol oxides in cooked chicken. In *Proceedings of the 41st Annual International Congress of Meat Technology*, pp. 372–373. (San Antonio: American Meat Science Association.

Gutteridge, J.M.C. and Halliwell, B. (1990a) The measurement and mechanism of lipid peroxidation in biological systems. *Trends in Biological Sciences*, **15**, 129–135.

Gutteridge, J.M.C. and Halliwell, B. (1990b) Inhibition of iron-catalysed formation of hydroxyl radicals from superoxide and of lipid peroxidation by desferrioxamine. *Biochemical Journal*, **184**, 469–472.

Halliwell, B. (1987) Free radicals and metal ions in health and disease. *Proceedings of the Nutrition Society*, **46**, 13–26.

Halliwell, B. and Gutteridge, J.M.C. (1986) Oxygen free radicals and iron in relation to biology and medicine. Some problems and concepts. *Archives of Biochemistry and Biophysics*, **246**, 501–514.

Harel, S. and Kanner, T. (1985) Muscle membranal lipid peroxidation initiated by H2O2-activated metmyoglobin. *Journal of Agricultural and Food Chemistry*, **33**, 1188–1192.

Hsieh, R.J. and Kinsella, J.E. (1989) Oxidation of PUFA: mechanisms, products and inhibition with emphasis on fish. *Advances in Food and Nutrition Research*, **33**, 233–341.

Huang, Y.X. and Miller, E.L. (1993) Iron-induced TBARS as an indicator of oxidative stability of fresh chicken meat. In *Proceedings of the 11th European Symposium on the Quality of Poultry Meat*, pp. 430–434. Edited by P. Colin, J. Culioli and F.H. Ricard. Tours: WPSA..

Kanner, J. and Doll, L. (1991) Ferritin in turkey tissue. A source of catalytic iron ions for lipid peroxidation. *Journal of Agricultural and Food Chemistry*, **39**, 247–249.

Kanner, J., Harel, S. and Granit, R. (1992) Oxidative processes in meat and meat products: quality implications. In *Proceedings of the 38th International Congress of Meat Science and Technology*, pp. 111–125. Clermont-Ferrand: ICOMST.

Kappus, H. (1991) Lipid peroxidation: mechanism and biological significance. In *Free Radicals and Food Additives*, pp. 59–75. Edited by O.I. Arouma and B. Halliwell. London: Taylor and Francis.

Kennedy, D.G., Goodall, E.A., McIlroy, S.G., Bruce, D.W. and Rice, D.A. (1991) The effects of increased vitamin E supplementation on profitable commercial broiler production. *Proceedings of the Nutrition Society*, **50**, 197A.

Lanari, M.C., Cassens, R.G., Schaefer, D.M. and Scheller, K.K. (1993) Dietary vitamin E enhances color and display life of frozen beef from Holstein steers. *Journal of Food Science*, **58**, 701–704.

Lavelle, C.L., Hunt, M.C. and Kropf, D.H. (1995) Display life and internal cooked colour of ground beef from vitamin E-supplemented steers. *Journal of Food Science*, **60**, 1175–1178.

Lin, C.F., Gray, J.I., Asghar, A., Buckley, D.J., Booren, A.M. and Flegal, C.J. (1989a) Effects of dietary oils and α-tocopherol supplementation on lipid composition and stability of broiler meat. *Journal of Food Science*, **54**, 1457–1460.

Lin, C.F., Gray, J.I., Asghar, A., Buckley, D.J., Booren, A.M., Crackel, R.L. and Flegal, C.J. (1989b) Effects of oxidized dietary oil and antioxidant supplementation on broiler growth and meat stability. *British Poultry Science*, **30**, 855–864.

Liu, Q., Lanari, M.C. and Schaefer, D.M. (1995) A review of dietary vitamin E supplementation for improvement of beef quality. *Journal of Animal Science*, **73**, 3131–3140.

Löliger, J. (1991) The use of antioxidants in foods. In *Free Radicals and Food Additives*, pp. 121–150. Edited by O.I. Arouma and B. Halliwell. London: Taylor and Francis.

Machlin, L.J. and Bendich, A. (1987) Free radical tissue damage: protective role of antioxidant nutrients. *FASEB Journal*, **1**, 441–445.

Marusich, W.L., DeRitter, E., Ogring, E.F., Keating, J., Mitrovic, M. and Bunnell, R.H. (1975) Effect of supplemental vitamin E in control of rancidity in poultry meat. *Poultry Science*, **54**, 831–844.

McIlroy, S.G., Goodall, E.A., Rice, D.A., McNulty, M.S. and Kennedy, D.G. (1993) Improved performance in commercial broiler flocks with subclinical infectious bursal disease when fed diets containing increased concentrations of vitamin E. *Avian Pathology*, **22**, 81–94.

McKay, P.B. and King, M.M. (1980) Vitamin E: its role as a biological free radical scavenger and its relationship to the microsomal mixed-function oxidase system. In *Vitamin E: a comprehensive treatise*, pp. 289–317. Edited by L.J. Machlin. New York, Marcel-Dekker.

Minotti, G. and Aust, S.D. (1992) Redox cycling of iron and lipid peroxidation. *Lipids*, **27**, 219–226.

Mitsumoto, M., Faustman, C., Cassens, R.G., Arnold, R.N., Schaefer, D.M. and Scheller, K.K. (1991) Vitamins E and C improve pigment and lipid stability in ground beef. *Journal of Food Science*, **56**, 194–197.

Mitsumoto, M., Arnold, R.N., Schaefer, D.M. and Cassens, R.G. (1993) Dietary versus postmortem supplementation of vitamin E on pigment and lipid stability in ground beef. *Journal of Animal Science*, **71**, 1812–1816.

Molenaar, I., Hulstaert, C.E. and Hardonk, M.J. (1980) Role in function and ultrastructure of cellular membranes. In *Vitamin E: a comprehensive treatise* pp. 474–494. Edited by L.J. Machlin. New York: Marcel-Dekker.

Monahan, F.J., Gray, J.I., Asghar, A., Buckley, D.J. and Morrissey, P.A. (1992a) Influence of dietary vitamin E (α-tocopherol) on the colour stability of pork chops. In *Proceedings of the 38th International Congress of Meat Science and Technology*, pp. 543–546. Clermont -Ferrand: ICOMST.

Monahan, F.J., Gray, J.I., Booren, A.M., Miller, E.R., Buckley, D.J., Morrissey, P.A. and Gomaa, E.A. (1992b) Influence of dietary treatment on lipid and cholesterol oxidation in pork. *Journal of Agricultural and Food Chemistry*, **40**, 1310–1315.

Monahan, F.J., Crackel, R.L., Gray, J.I., Buckley, D.J. and Morrissey, P.A. (1993a) Catalysis of lipid oxidation in muscle model systems by haem and inorganic iron. *Meat Science*, **34**, 95–106.

Monahan, F.J., Buckley, D.J., Morrissey, P.A. and Lynch, P.B. (1993b) Effect of dietary α-tocopherol acetate on the α-tocopherol levels in porcine muscle and on lipid oxidation in pork. In *Safety and quality of food from animals, Occasional publication No. 17*, pp. 104–107. Edited by D.J. Wood and T.J.L. Lawrence. British Society of Animal Production.

Monahan, F.J., Gray, J.I., Asghar, A., Haug, A., Strasburg, G.M., Buckley, D.J. and Morrissey, P.A. (1994) Influence of diet on lipid oxidation and membrane structure in porcine muscle microsomes. *Journal of Agricultural and Food Chemistry*, **42**, 59–63.

Morrissey, P.A. and Tichivangana, J.Z. (1985) The antioxidant activities of nitrite and nitrosylmyoglobin in cooked meats. *Meat Science*, **14**, 175–190.

Morrissey, P.A., Buckley, D.J., Sheehy, P.J.A. and Monahan, F.J. (1994) Vitamin E and meat quality. *Proceedings of the Nutrition Society*, **53**, 289–295.

Mueller, D.P.R., Manning, J.A., Mathias, P.M. and Harries, J.T. (1976) Studies on the intestinal hydrolysis of tocopheryl esters. *International Journal of Vitamin and Nutrition Research*, **46**, 207–210.

National Research Council (1984) *Nutrient Requirements of Beef Cattle (6th edition)*. Committee on Animal Nutrition, Board on Agriculture. Washington D.C.: National Academy of Sciences.

National Research Council (1988) *Nutrient Requirements of Swine (9th edition)*. Committee on Animal Nutrition, Board on Agriculture. Washington D.C.: National Academy of Sciences.

National Research Council (1989) *Recommended Dietary Allowances*. Committee on Dietary Allowances, Food and Nutrition Board. Washington D.C.: National Academy of Sciences.

National Research Council (1993) *Nutrient Requirements of Fish*. Committee on Animal Nutrition, Board on Agriculture. Washington D.C.: National Academy of Sciences.

National Research Council (1994) *Nutrient Requirements of Poultry (9th edition)*. Committee on Animal Nutrition, Board on Agriculture. Washington D.C.: National Academy of Sciences.

O' Neill, L., Morrissey, P.A. and Buckley, D.J. (1995) Effect of dietary fat and vitamin E supplementation on the oxidative stability and sensory quality of chicken muscle. *Irish Journal of Agricultural and Food Research*, **34**, 205–206A.

Ristic, M. (1991) Einfluss der Vitamin-E-Versorgung auf die Fleischqualitaet von Broilern. *Mitteilungsblatt der Bundesanstalt fuer Fleischforschung, Kulmbach,* **111**, 11–13.

Sato, K. and Hegarty, G.R. (1971) Warmed-over flavour in cooked meats. *Journal of Food Science,* **36**, 1098–1102.

Schaefer, D.M., Liu, Q., Faustman, C. and Yin, M-E. (1996) Supranutritional administration of vitamins E and C improves oxidative stability of beef. *Journal of Nutrition,* **125**, 1792S–1798S.

Schaich, K.M. (1992) Metals and lipid oxidation. Contemporary issues. *Lipids,* **27**, 209–218.

Sheehy, P.J.A., Morrissey, P.A. and Flynn, A. (1990) Effect of dietary α-tocopherol level on susceptibility of chicken tissues to lipid peroxidation. *Proceedings of the Nutrition Society,* **49**, 28A.

Sheehy, P.J.A., Morrissey, P.A. and Flynn, A. (1991) Influence of dietary α-tocopherol on tocopherol concentrations in chick tissues. *British Poultry Science,* **32**, 391–397.

Sheehy, P.J.A., Morrissey, P.A. and Flynn, A. (1993a) Increased storage stability of chicken muscle by dietary α-tocopherol supplementation. *Irish Journal of Agricultural and Food Research,* **32**, 67–73.

Sheehy, P.J.A., Morrissey, P.A. and Flynn, A. (1993b) Consumption of thermally oxidized sunflower oil by chicks reduces α-tocopherol status and increases susceptibility of tissues to lipid oxidation. *British Journal of Nutrition,* **71**, 53–65.

Sheehy, P.J.A., Morrissey, P.A. and Flynn, A. (1993c) Influence of heated vegetable oils and α-tocopheryl acetate supplementation on α-tocopherol, fatty acids and lipid peroxidation in chicken muscle. *British Poultry Science,* **34**, 367–381.

Sheldon, B.W. (1984) Effect of dietary tocopherol on the oxidative stability of turkey meat. *Poultry Science,* **63**, 673–681.

Sklan, D., Bartov, I. and Hurwitz, S. (1982) Tocopherol absorption and metabolism in the chick and turkey. *Journal of Nutrition,* **112**, 1394–1400.

Sklan, D., Tenne, Z. and Budowski, P. (1983) The effect of dietary fat and tocopherol on lipolysis and oxidation in turkey meat stored at different temperatures. *Poultry Science,* **62**, 2017–2021.

Steinberg, D., Parthasarathy, S., Caron, T.E., Khoo, J.C. and Witztun, J.L. (1989) Modification of low density lipoprotein that increases its atherogenecity. *New England Journal of Medicine,* **320**, 915–924.

Tichivangana, J.Z. and Morrissey, P.A. (1985) Metmyoglobin and inorganic metals as pro-oxidants in raw and cooked muscle systems. *Meat Science,* **15**, 107–116.

Ward, N.E. (1993) Vitamin supplementation rates for US commercial broilers, turkeys, and layers. *Journal of Applied Poultry Research*, **2**, 286–296.

Warkup, C. (1994) Vitamin E and pigmeat quality. In *Proceedings of Symposium on Vitamin E and Meat Quality*, pp. 1–16. Bologna: Istituto delle Vitamine.

Wen, J., Morrissey, P.A., Buckley, D.J. and Sheehy, P.J.A. (1996) Oxidative stability and α-tocopherol retention in turkey burgers during refrigerated and frozen storage as influenced by dietary α-tocopheryl acetate. *British Poultry Science*, **37**, 787–795.

Williams, S.N., Frye, T.M., Frigg, M., Schaefer, D.M., Scheller, K.K. and Liu, Q. (1992) Vitamin E as an *in situ* post-mortem pigment and lipid stabiliser in beef. In *Proceedings of the Pacific Northwest Animal Nutrition Conference*. pp. 149. Spokane.

2

NUTRITIONAL MANIPULATION OF IMMUNE COMPETENCE IN YOUNG NON-RUMINANT ANIMALS

A. PUSZTAI[a], E. GELENCSÉR[b], G. GRANT[a] and S. BARDOCZ[a]

[a]Rowett Research Institute, Bucksburn, Aberdeen AB21 9SB
[b]Central Food Research Institute, Budapest 114, P.O. Box 3939, H-1537, Hungary

Introduction

It is a common experience that, when plant proteins are used in the diet of young animals, nutritional disturbances of varying severity can occur such as:

* poor nutritional performance
* reduced growth
* reduced digestion/absorption
* changes in gut motility
* structural damage in the small intestine
* diarrhoea

One of the most studied examples of these is the adverse reaction that occurs with preruminant calves after weaning them on to steam-heated soya bean milk replacer diets (Sissons, 1989; Lallés, 1993). However, preruminant calves are by no means the only animals affected. Indeed, most young animals experience some form of digestive disturbances of varying duration after an early exposure to plant proteins in their diet. Although these damaging and costly responses to plant proteins have been known for a long time, their elimination has proved to be elusive mainly because the reaction mechanism of the disturbances is poorly understood.

Most legume seed meals contain a variety of antinutrients and these are known to have a negative effect on nutritional performance (Huisman and Jansman, 1991). However, protein antinutrients, such as lectins, digestive enzyme inhibitors and other deleterious proteins, are usually efficiently inactivated by proper heat-treatment in most common feedstuffs. It is therefore unlikely that the observed damaging effects in young animals are due to the presence of antinutrients. Although there can be many other possible reasons for the digestive disturbances, it is likely that humoral and/or cellular immune responses, gut mucosal and systemic hypersensitivity and allergic reactions to plant protein antigens, singly or in combinations, may contribute to the overall nutritional damage. Although seed proteins of nutritional importance, particularly those of low molecular weight, are powerful antigens (Hessing, Bleeker, van Biert, Vlooswijk and van Oort, 1994), most seed proteins are denatured by heat treatment, become highly digestible and lose their native antigenicity. However, as low molecular weight proteins are more resistant to heat, the antigenic response of young animals to the heat-treated meal is likely to be different and in some cases enhanced, in comparison with the original raw meal. Thus, as a result of these changes in protein antigenicity/ allergenicity, powerful immune responses can be expected to occur, particularly in the immature gut of young animals.

In the first part of this review the likely causes of the damaging reactions to plant proteins and possible strategies and treatments for their elimination will be discussed with particular emphasis on systemic and mucosal immune responses. In the second part an example is given of how dietary manipulation can be used to minimise the consequences of potentially harmful gut reactions to untreated, raw soya whey of high lectin content. The success of this strategy has been shown to be dependent on the presence of functionally active soya lectin in the diet (Pusztai, Grant, Bardocz, Gelencsér and Hajos, 1997). Although its effects on the gut were likely to be complex, it is possible that high mucosal sIgA and IgM responses to the lectin and possibly to other proteins resulting from dietary exposure could have prevented the systemic absorption of soya antigens and thus contributed to the improvement in the nutritional value of the diet.

Approaches for the elimination of digestive disturbances

METHODS TO REDUCE ANTIGENICITY OF PLANT PROTEINS

The usual technological approach to make proteins in the diet more palatable and digestible is to use various forms of heat treatment. Steam-heating has been one of the most favoured methods of industry particularly for the inactivation of lectins and trypsin inhibitors (Huisman and Jansman, 1991). However, other treatments such as extrusion, microwaving, enzymic pre-digestion, hot aqueous ethanol extraction and exposure to alkaline and/or acidic conditions can also reduce the antigenicity of seed meals and allow higher inclusion levels of seed meal proteins in milk replacer formulas (Lallés, 1993). Unfortunately, none of these technological treatments eliminate short-term or long-term digestive disturbances in young animals. Indeed, for reasons explained in the Introduction, the antigenicity of proteins in the meals is usually only modified by these treatments but not eliminated. Moreover, the changes introduced by the various treatments are likely to result in different products in a rather unpredictable way, adding more uncertainty to the outcome of the feeding in practice.

Repair of structural damage to gut mucosa

One of the consequences of feeding young animals on diets containing antigenic seed proteins particularly after weaning is the occurrence, in addition to digestive disturbances, of structural damage to the gut mucosa, including villus atrophy and crypt hyperplasia (Pedersen and Sissons, 1984; Seegraber and Morrill, 1986). This small intestinal damage is reversed quickly after the animals are returned to weaning diets containing milk proteins. Although polyamine supplementation has also been suggested to have a beneficial effect on the repair of the damage (Grant, Thomas, King and Liesman, 1990), most of the benefits of polyamine supplementation are marginal. In fact, the only sure way to eliminate the disturbances is to withdraw plant proteins from the diet, strongly suggesting that the damaging effects are most likely to be due to immune responses to these proteins in the diet.

Immune-related strategies affecting the systemic immune system

It is clear that both the occurrence of a powerful systemic immune response and the lack of an adequate response by the mucosal secretory immune system to orally presented antigens are equally deleterious. Indeed, animals are best able to avoid or minimise digestive disturbances when their response to dietary antigens leads simultaneously to systemic tolerance and a high mucosal secretory response. Unfortunately, this is very seldom, if ever, achieved. There are several reasons for this. One of the main problems is that orally presented antigens usually affect both systemic and mucosal immune systems. Although the responses of the two systems are linked, this linkage is not easily predictable and, with a few exceptions, does not generally work to the advantage of nutritionists. Furthermore, although the systemic immune system can be stimulated to respond to antigen challenge, particularly when administration is by the parenteral route, the results of oral immunisation are not clear-cut. An even more serious problem is that mucosal secretory responses to food antigens are usually poor.

It is commonly accepted that the presence of food antigens in the gut lumen, particularly in young animals, affects systemic immunity (Chase, 1946). Thus, dietary antigens either prime/immunise the systemic immune system or induce hyporesponsiveness/tolerance in it (Challacombe and Tomasi, 1987). Both humoral and cell-mediated responses are affected. One of the main factors influencing whether oral antigens prime or induce tolerance in the systemic immune system is the nature of the antigen. Thus, luminal soluble antigens, particularly when applied at low concentrations, have a tendency to induce systemic tolerance while particulate antigens or soluble antigens which easily aggregate usually prime the systemic immune system (Challacombe and Tomasi, 1987). When particulate antigens such as bacterial cell walls induce systemic tolerance, this requires the administration of large amounts of antigen and the tolerance is relatively short-lived. It is also important whether the antigen is a protein or polysaccharide and whether it is easily digestible in the alimentary tract. The antigen dose, frequency of administration and duration of exposure are all important determinants of whether priming or tolerance is induced. For example, feeding mice with very small or very large doses of ovalbumin leads to tolerance while moderate doses result in priming. In most instances the immune status and, particularly prior antigen exposure, materially affects the outcome. Thus, if systemic exposure to an antigen is preceded by intragastric immunisation, the usual result is tolerance. However, if intraperitoneal immunisation is followed by intragastric exposure to the same antigen, the humoral IgG response is boosted (Challacombe and Tomasi, 1987). It is also commonly observed that tolerance is more easily induced in some animals than others and the age of the animal influences the response although not to the same extent as found with secretory responses. Thus, although tolerance to ovalbumin can be induced in young animals, it is less extensive than in adults. Finally, the gut flora and presence of LPS (bacterial lipopolysaccharide) are also major factors which help the inducement of oral tolerance in animals. Although it is not clear how these factors determine the eventual response, it is known that oral tolerance is mediated by various humoral factors including serum suppressive factors, with possible influences by anti-idiotypic antibodies, while the most important cellular factors involved are the suppressor cells.

Tolerance is known to occur in practical situations, a classical example occurring in piglets exposed via the diet to soya antigens prior to weaning (Stokes, Miller and Bourne, 1987). Thus, in one experiment, piglets were divided into three groups prior to weaning. Piglets in the first group received no feed at all (naive animal group), those in the second group were given small amounts of soya feed (primed group), while those in the third group were exposed to large amounts of postweaning soya feed (tolerized group). When the piglets were weaned at 3 weeks on to soya-based postweaning diet, diarrhoea and reduction in xylose absorption were greatest in the primed group and intermediate in naive animals. In contrast, diarrhoea was nearly absent and xylose absorption was only slightly affected in tolerized piglets. However, if the piglets were not weaned or weaned on to non-antigenic diet, such as pre-digested milk protein, no nutritional disturbances were found. The demonstration of these beneficial effects of preweaning exposure to the weaning diet in this experiment was impressive and underlined the general feeling that oral tolerance as a second line defence is desirable. However, the difficulty in achieving the desired result in practice in a predictable way somewhat detracts from the value of this approach, particularly because it is difficult to control the dietary intake of the piglets before weaning. Moreover, the results of these experiments are also complicated by the undoubted occurrence of mucosal damage, postweaning allergic and hypersensitivity reactions. Another major drawback of this method of oral tolerization is that high oral tolerance may lead to secretory unresponsiveness and vice versa. Finally, it is also known that oral tolerance may be abrogated in pathological conditions of the gut.

Immune-related strategies affecting the mucosal secretory immune system

There is general agreement that increased mucosal secretory antibody response as a first line of defence against pathogens/allergens is desirable because it may lead to immune exclusion and prevention of food antigens reaching the systemic circulation (Challacombe, 1987). Unfortunately, there are difficulties in raising immunity against orally administered soluble antigens/allergens and mucosal immunity is less well understood than the systemic immune system. However, one of the advantages of mucosal immunity is that in some circumstances its stimulation is independent of the systemic immune system. This is because the mucosal immune system functions locally using antigen presenting cells, T and B cells which are all present in Peyer's patches that are specialized immune regions of the gastrointestinal tract (Challacombe, 1987). The functional immunoglobulin here is sIgA, a specific secretory antibody found in all mucosal tissues including the gut.

Induction of a secretory antibody response to orally ingested antigens is dependent on the nature and dose of the antigen, whether it is soluble or particulate, the frequency and duration of oral antigen challenge, the site of immunisation and the age and species of host animal. In most instances particulate antigens are more efficient than soluble one in raising a secretory response but even then very large oral doses are needed. In man the secretory sIgA system appears to mature faster than the systemic immune system. However, in mice apparently the reverse is true (Challacombe, 1987). Previous exposure of the host to antigen also has a major influence on the success of mucosal immunisation, although the direction of the response is ambivalent. Thus, lambs were protected against *Salmonella typhimurium* infection by a combination of systemic priming and oral boosting

(Husband, 1978). However, in many instances prior systemic exposure to the same antigen tends to inhibit the mucosal response (Pierce and Koster, 1983) while repeated mucosal exposures boost it. The most important factors controlling the success of this mucosal secretory response are whether vaccination is carried out with living or dead organisms and, particularly, whether adjuvants are used or not. A-B toxins, such as cholera toxin (CT) and *E. coli* heat-labile toxin (LT), both of which contain a ribosome-inactivating A subunit and a pentameric arrangement of lectinic B subunits, are particularly effective in raising powerful sIgA responses (Challacombe, 1987; Lycke, Bromander & Holmgren, 1989). Unfortunately, pathogens not producing A-B toxins (Salmonella, etc) and the non-toxic B subunits of CT or LT are less effective adjuvants unless co-applied with adjuvant amounts of the full toxin (Figure 2.1). Other oral agents such as muramyl dipeptide, polycationic DEAE-dextran and liposomes, particularly in the presence of muramyl dipeptide, are also effective adjuvants to boost secretory responses.

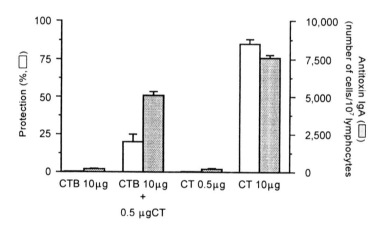

Figure 2.1 Protection is estimated by measuring fluid accumulation in intestinal loops of immunised *vs.* non-immunised mice and antitoxin-IgA lamina propria cells were counted in immunised mice; values in the bar diagram were taken from Lycke, Bromander and Holmgren, 1989; by permission).

Although this strategy appears to be promising particularly because it may be possible to raise the sIgA titre against soluble antigens when they are co-administered with adjuvant amounts of CT or LT, it has not been fully tested and may be hazardous in practice. Additionally, oral administration of CT and LT unfortunately also primes the systemic immune system and no oral tolerance is obtained. Moreover, it is not always predictable whether systemic pre-exposure increases or prevents the development of sIgA to the same antigens.

Mucosal cellular immune responses to orally administered antigens also occur but they are far less well known than the secretory responses. Cytotoxic T cells can be present in Peyer's patches and intraepithelial lymphocyte infiltration and graft-versus-host disease can occur but it is not clear how they affect cellular responses.

Novel dietary strategy to boost sIgA response and reduce antinutrient effects

Feeding young animals diets based on soya bean meal presents particular difficulties. This is because the same heat treatment that is used to inactivate the lectin and two protease inhibitor antinutritive components of soya, also induces poorly defined changes in the antigenicity of soya proteins. By common experience, the incorporation of these processed soya products in animal feeds can lead to damaging immune responses. The usual approach for minimising these problems is to include soya products in the diet at low levels based on the expectation that their harmful effects are diluted out. However, this strategy is often wasteful.

In a novel approach the question was asked whether it would be possible to raise a significant mucosal secretory sIgA response to soya proteins in the diet of young rats and thus protect them from the potentially harmful antigenic/allergenic effects of the feed ? As CT and LT and, to a lesser extent their non-toxic lectinic B subunit, were excellent adjuvants for raising powerful mucosal sIgA responses not only to the toxins but also to other proteins co-applied with them, it was hoped that the undenatured and functionally active agglutinin in raw soya diets could also have a similar and significant adjuvant effect. Furthermore, if this could be achieved while still retaining the benefits of gut stimulation by the soya lectin, a dietary strategy incorporating these two aspects of lectin effects could be of particular value in animal nutrition. As is described below, by feeding rats in a novel strategy of short cycles in which the protein source of the diet alternated between unprocessed raw soya albumin (whey) of high lectin and trypsin inhibitor content and a high-quality protein (e.g. lactalbumin), it was indeed possible to induce a highly significant mucosal secretory antibody response to the soya agglutinin and also to improve the nutritional utilisation of the antinutrient-rich soya albumins.

EXPERIMENTAL DESIGN

In these experiments groups of male Hooded-Lister rats (Rowett strain) weaned at 19 days were used as previously described (Pusztai *et al.*, 1997). The rats were divided into 2 groups, 5 rats in each group. The diet for the experimental group contained 100 g protein/kg diet based on a soya albumin fraction, SBALB. This preparation contained all the proteins remaining soluble after removing globulins by isoelectric precipitation from aqueous extracts of defatted soya meal. This fraction is rich in SBA (40-60 g/kg) and the two trypsin inhibitors (up to 300 g/kg). The control group of rats were fed LA diet (100 g lactalbumin/kg diet) throughout the experiment and their intake was restricted to the voluntary intake of the test rats. The experimental design was such that initially the soya group was fed soya diet for 7 days, switched to LA diet for 8 days, followed by soya diet for 7 days and then LA diet for a further 7 days. Next, after another 6 days on soya diet followed by 20 days on LA diet, the rats were finally exposed to soya diet for a 5 day period and killed by halothane overdose on the 61th day of the combined feeding experiment and dissected. All tissues and bodies were freeze-dried, weighed and analysed for the determination of protein and lipid contents. Throughout the experiment faeces were collected daily and used for nitrogen determinations and the estimation of specific faecal anti-SBA antibodies. A blood sample was also taken at each stage from the tail vein of rats and used for the measurement of serum anti-SBA IgG. A control experiment was also carried out (results are not described but see Pusztai *et al.*, 1997) in which the rats were first fed a diet containing an SBALB preparation from which the lectin had been selectively removed and then switched to the LA diet as in the main experiment described above. The results were then subjected to statistical analyses by one-way ANOVA using the 'Minitab' computer programme (Penn State University, State College, PA, U.S.A.). The significance of difference between treatment groups was estimated using Student's *t* test.

Nutritional benefits and immune responses

The growth of the test rats was always depressed in the lectin part of the feeding cycles. Despite this, when the comparisons were made at the end of the LA feeding part of the cycles, growth of the test rats became comparable to that of control rats pair-fed on high-quality LA diet throughout (Figure 2.2).

The presence of SBA reduced the efficiency of feed conversion in the test group in the antinutritive phase of the cycle. However, this was not because the soya diet was poorly digested. Indeed, except for the first dietary exposure to SBA there was no significant difference in the digestibility of the diets by the two groups of rats (Figure 2.3), particularly taking into account that most of SBA escapes digestion in the alimentary tract of the rats (Figure 2.4; and Pusztai *et al.*, 1997).

However, the rats compensated for the poor feed conversion of the antinutritive phase in the following benefit phase because in this part of the cycle they utilised the LA diet with increased efficiency in comparison with the control group (Table 2.1). As with the selective removal of SBA from SBALB most of the gains in the

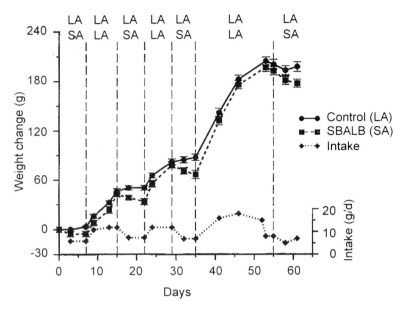

Figure 2.2 Weight changes of test rats fed in short cycles of diets in which the protein source alternated between soya albumin proteins and lactalbumin in comparison with control rats pair-fed lactalbumin diet throughout. Dietary intakes are also given in the diagram (from Pusztai *et al.*, 1997; by permission of Cambridge University Press).

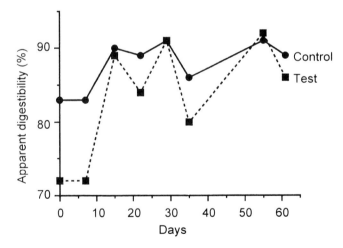

Figure 2.3 Changes in apparent digestibility of test and control rats during the diet-switching experiment.

benefit part of the feeding cycle were lost (results not given here but are described in Pusztai *et al.*, 1997), it is clear that for the success of this strategy it is essential that the diet should contain the SBA lectin in a fully functionally active state and in a form that enables it to bind to the gut surface.

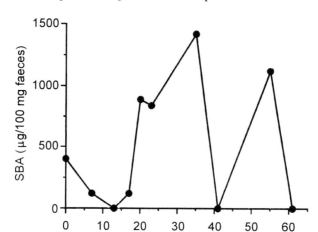

Figure 2.4 Amounts of soya agglutinin, SBA, in faeces of test rats during the experiment estimated by competitive indirect ELISA using specific anti-SBA rabbit antibody.

Table 2.1 WEIGHT CHANGES AND FEED CONVERSION EFFICIENCY OF RATS IN TEST AND CONTROL PERIODS RESPECTIVELY DURING THE EXPERIMENT

Treatment	Test period Soya or LA diets			Control Period (LA diet)		
	Initial weight (g)	Final weight (g)	Feed conversion (g/g intake)	Initial weight (g)	Final weight (g)	Feed conversion (g/g intake)
Switch 1 1-7d				8-15d		
Control	88.8±3.5[a]	92.2±1.0[a]	0.08±0.02[a]	92.2±1.0[a]	133.0±3.0[a]	0.43±0.03[a]
Test	86.6±2.1[a]	81.7±3.2[b]	negative[b]	81.7±3.2[b]	129.0±3.0[a]	0.49±0.03[b]
Switch 2 16-22d				23-29d		
Control	133.0±3.0[a]	137.0±2.8[a]	0.06±0.04[a]	137.0±2.8[a]	168.5±3.7[a]	0.38±0.04[a]
Test	129.0±3.0[a]	120.0±3.9[b]	negative[b]	120.0±3.9[b]	163.5±3.7[a]	0.52±0.04[b]
Switch 3 30-35d				36-55d		
Control	168.5±3.7[a]	174.0±4.0[a]	0.11±0.08[a]	174.0±4.0[a]	282.5±7.5[a]	0.36±0.02[a]
Test	163.5±3.7[a]	153.0±5.0[b]	negative[b]	153.0±5.0[b]	281.6±5.1[a]	0.43±0.02[b]
Switch 4 56-61d						
Control	282.5±6.0[a]	283.5±6.0[a]	0.01±0.08[a]			
Soya albumin	281.6±5.1[a]	263.6±7.2[b]	negative[b]			

Results are means ± SD for 5 rats per group. For each period values in a vertical row with distinct supercripts differ significantly (P≤0.05)
(From Pusztai *et al.*, 1997; by permission of Cambridge University Press).

Although the mechanism of SBA action is not clear it is possible that some of the known interactions of lectins with the gut epithelium may potentially contribute to the final outcome. Thus, due to the hyperplastic growth of the gut, particularly the small intestine (Bardocz, Grant, Ewen, Duguid, Brown, Englyst and Pusztai, 1995; Bardocz, Grant, Franklin, Pusztai and Carvalho, 1996) and its full reversibility, proteins accumulating in gut tissues in the antinutritive phase may be made available for body metabolism in the benefit phase. As at the same time the gut returns to pre-stimulation size, the nutritional cost of its turnover is reduced and more of the protein and energy in the diet can be utilised by the body. Furthermore, as nutrient absorption through the freshly renewed gut epithelium is also improved in the benefit phase, all the factors favouring efficient nutrient utilisation are present and fully operational to achieve maximum efficiency.

Further improvement in the nutritional performance of the rats may have resulted from the stimulation by dietary SBA of highly significant and specific anti-SBA sIgA and IgM responses in the gut (Figure 2.5). Possible protection by these mucosal antibodies may then be due to their immune exclusion potency and blockage of the systemic absorption of functionally reactive SBA into the blood circulation and thus the inhibition of unwanted and harmful systemic metabolic effects and immune responses. Indeed, the finding of relatively low systemic anti-SBA IgG levels in rat sera (Figure 2.6) is probably a true reflection of the low systemic absorption of SBA. Although not described here in detail, mucosal secretory antibody responses were not confined to anti-SBA sIgA but also included a significant response to the two trypsin inhibitors. The possibility exists that, similar to the adjuvant effect of A-B type toxins, SBA may have had significant adjuvant effect for raising mucosal responses to other soya antigens in the diet and thus inhibited their absorption from the gut lumen into the blood circulation and reduced potentially harmful systemic reactions to them.

Conclusion

It is clear from the short literary summary given here of the possible responses to dietary antigens of both systemic and mucosal immune systems that these are highly complex reactions. Indeed, they would have been even more complex had the immediate (IgE-mediated) and cell-mediated delayed allergenic responses also been considered. Although these reactions clearly occur at the gut level (Brandtzaeg, 1987), most of the IgE is not produced locally in the gut mucosa but in regional lymph nodes and the spleen and are only recruited by the intestinal mast cells. The only well-studied exception is the copious IgE production in the gut after infestation by a gut-dwelling helminth parasite (*Nippostrongylus brasiliensis*). Therefore in a nutritional context, the importance of systemic

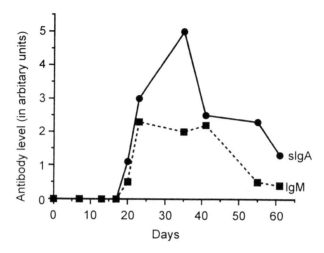

Figure 2.5 Specific anti-SBA sIgA and IgM levels in faeces of test rats during the diet-switching experiment determined by ELISA using anti-rat IgA and IgM second antibodies.

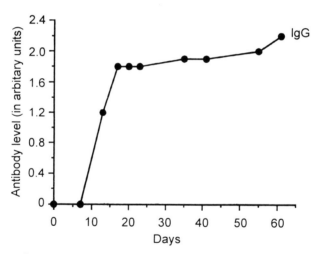

Figure 2.6 Changes in serum IgG levels of rats reacting specifically with soya agglutinin, SBA, determined by ELISA

tolerance and high mucosal secretory response to dietary antigens may override all others. Unfortunately, these two apparently beneficial responses seldom occur concurrently. Thus, unless CT or LT are used as adjuvants, the oral administration of soluble antigens is not very effective in raising mucosal secretory immunity although soluble antigens can be good inducers of oral tolerance. In contrast, particulate antigens that are moderately good stimulants of sIgA responses in the gut, do not induce oral tolerance in most instances. Both immune systems can respond to dietary antigens in unison or independent of one another. As these reactions can be differently affected by practically everything that concerns the antigen and the host, including the age and species of the animals, the nature of the antigens, previous exposures to them and the frequency of administration, the overall response in most instances is unpredictable. Although the beneficial effect of weaning young animals to non-antigenic diets is undoubtedly true, most proteins of non-maternal origin are "foreign" antigens and therefore the exclusion of antigenic activity from the diet under practical feeding conditions may be very difficult to achieve. Delaying the time of weaning may also reduce harmful immune responses. However, this strategy has many practical drawbacks. Thus, it is difficult at the moment to draw up generally applicable methods of dietary manipulation which would guarantee a successful outcome and therefore any possible new strategies aimed at lessening the impact of exposing young animals to foreign antigens in the diet need first to be fully tested in model experiments. Indeed, although the benefits obtained by the strategy described in the second part of this review are tangible, they cannot be generalised, particularly if any attempt is made to use this strategy with preruminant animals. Although the cyclic use of a fully functionally active lectin such as SBA in the diet had clear nutritional benefits and stimulated a major mucosal secretory response to SBA and other soya antigens, there is as yet no evidence of a causal link between these two events. Although this may be of less concern for practical nutritionists, the possible unequivocal scientific demonstration that the two events are interlinked, particularly if it is established how this occurs, should take the significance of these findings beyond practical empiricism and extend them to general validity and use. Additionally, this method may offer an alternative and cost-effective use of potential waste products such as soya whey.

References

Bardocz, S., Grant, G., Ewen, S.W.B., Duguid, T.J., Brown, D.S., Englyst, K. and Pusztai, A. (1995) Reversible effect of phytohaemagglutinin on the groth and metabolism of rat gastrointestinal tract. *Gut*, **37**, 353–360.

Bardocz, S., Grant, G., Franklin, M.F., Pusztai, A. and Carvalho, A. de F.F.U. (1996) The effect of phytohaemagglutinin on the growth, body composition and plasma   insulin of the rat at different dietary concentrations. *British Journal of Nutrition*, **76**, 613–626.

Brandtzaeg, P. (1987) The B cell system. In *Food Allergy and Intolerance*. Chapter 7, pp 118–155. Edited by J. Brostoff and S.J. Challacombe. Bailliére Tindall: London.

Challacombe, S.J. (1987) The induction of secretory IgA responses. In *Food Allergy and Intolerance*. Chapter 15, pp 269–285. Edited by J. Brostoff and S.J.Challacombe.Bailliére Tindall: London.

Challacombe, S.J. and Tomasi, T.B. (1987) Oral tolerance. In *Food Allergy and Intolerance*.Chapter 14, pp 255–268. Edited by J.Brostoff and S.J. Challacombe. Bailliére Tindall: London.

Chase, M.W. (1946) Inhibition of experimental drug allergy by prior feeding of the sensitizing agent. *Proceedings of the Society for Experimental Biology and Medicine*. **61**, 257–259

Grant, A.L., Thomas, J.W., King, K.J. and Liesman, J.S. (1990) Effects of dietary amines on small intestine variables in neonatal pigs fed soy protein isolate. *Journal of Animal Science*, **68**, 373–381.

Hessing, M., Bleeker, H., van Biert, M., Vlooswijk, R.A.A. and van Oort, M.G. (1994) Antigenicity of legume proteins. *Food and Agricultural Immunology*, **6**, 315–320.

Huisman, J. and Jansman, A.J.M. (1991) Dietary effects and some analytical aspects of antinutritional factors in peas (*Pisum sativum*), common beans (*Phaseolus vulgaris*) and soyabeans (*Glycine max*) in monogastric farm animals. *Nutritional Abstracts and Reviews* (series B), **61**, 901–921.

Husband, A.J. (1978) An immunisation model for the control of infectious enteritis. *Research in Veterinary Science*, **25**, 173–177.

Lallés, J.P (1993) Nutritional and antinutritional aspects of soyabean and field pea proteins used in veal calf production, a review. *Livestock Production Science*, **34**, 181–202.

Lycke, N., Bromander, A.K. and Holmgren, J. (1989) Role of local IgA antitoxin-producing cells for intestinal protection against cholera toxin challenge. *International Archives of Allergy and Applied Immunology*, **88**, 273–279.

Pedersen, H.E. and Sissons, J.W. (1984) Effect of antigenic soyabean protein on the physiology and morphology of the gut in the preruminant calf. *Canadian Journal of Animal Science*, **64**, 183–184.

Pierce, N.F. and Koster, F.T. (1983) Parenteral immunisation causes antigen-specific cell-mediated suppression of an intestinal IgA response. *Journal of Immunology*, **131**, 115–119.

Pusztai, A., Grant, G., Bardocz, S., Gelencsér, E. and Hajos, Gy. (1997) Novel dietary strategy for overcoming the antinutritional effects of soya whey of high agglutinin content. *British Journal of Nutrition*, in press.

Seegraber, F.J. and Morrill, J.L.(1986) Effect of protein source in calf milk replacers on morphology and absorptive ability of the small intestine. *Journal of Dairy Science*,**69**, 460–469.

Sissons, J.W. (1989) Aetilogy of diarrhoea in pigs and preruminant calves. In *Recent Advances in Animal Nutrition - 1989*, pp 261–282. Edited by W. Haresign and D.J.A. Cole, Butterworths Scientific Press, London.

Stokes, C.R., Miller, B.G. and Bourne, F.J. (1987) Animal models of food sensitivity. In *Food Allergy and Intolerance*. Chapter 16, pp 286–300. Edited by J. Brostoff and S.J. Challacombe. Bailliére Tindall: London.

3

ALTERNATIVES TO ANTIBIOTICS - THE INFLUENCE OF NEW FEEDING STRATEGIES FOR PIGS ON BIOLOGY AND PERFORMANCE

L. GÖRANSSON
Lantmännen Feed Development, Swedish Pig Center, Pl 2080,26800 Svalöv - *Sweden*

Introduction

Feed antibiotics (i.e. growth promoters) were banned in Sweden by law in 1986. The decision of the authorities followed a proposal from the Federation of Swedish Farmers based on concern of the consumers. At present the concern in question is the risk of cross resistance between growth promoters and therapeutic human antibiotics. This topic was reviewed by Lange and Ek (1995).

According to Robertsson (1994) the most significant consequence of prohibiting the use of growth promoters in pig production was a noticable reduction in post weaning daily weight gain and a twofold increase in post weaning diarrhoea (PWD).

During the first year of the ban, the veterinary prescription of Bayonox[R], the most commonly employed PWD drug in Sweden, was very low. From 1987, however, a reasonable amount of antibiotics was prescribed. Statistics from the feed manufacturers revealed considerable regional differences with respect to veterinary prescription of medicated diets. Despite the increase in the prescribed amount of olaquindox, the total amount in 1993 was only some 50% of the amount used as general feed additive in 1985 (Björneroth et al.,1995). The recommended therapeutic dose of olaquindox is 100 - 160 mg per kg dry feed compared to the 50 mg included as a feed additive before 1986. Consequently less than 20% of the piglets produced in Sweden in 1993 were treated therapeutically with olaquindox in their post-weaning diets. With respect to growing finishing pigs, no detrimental effects have emerged following taking the feed antibiotics out of the diet.

The pathogenesis of post weaning scouring is complex. Background factors which act alone or in combination are feed-feeding, environment-management and genetics. As the post weaning period has been the most critical phase during which detrimental effects following the exclusion of growth promoters, this presentation will deal mainly with this period of time.

Feed

PROTEIN

According to experience gained from practical pig units, the risk of PWD increases with the crude protein content of the creep feed. This has been demonstrated experimentally by Prohaszka and Baron (1980) and Danielsen (1984). It is possible however that the diet protein content *per se* is not the only predisposing factor, since allergic effects of some proteins have also been demonstrated (Miller *et al.* 1983) .

FIBRE

Dietary fibre, defined as non starch polysaccharides, constitutes a large group of compounds. The intestinal properties of different dietary fibres demonstrate great variability. Model experiments with the loop technique demonstrated a lowered risk of diarrhoea in pigs given beet pulp fibre (Larsen, 1981). Reduced length and severity of PWD has been reported as a direct effect of increased crude fibre content of the diet (Ball & Aherne 1987). On-farm production trials support these findings by demonstrating a lower PWD incidence in piglets given plantago polysaccharides or beet pulp fibre in the diet (Table 3.1 and 3.2, Göransson *et al.*,1995).

ANTI SECRETORY PROTEIN (ASP)

ASP is a regulatory protein which inhibits pathological intestinal fluid secretion induced by entero-toxins (Lange &Lönnroth, 1984; Lönnroth & Lange, 1986). ASP seems to have an important role in the defence against diarrhoea diseases (Lange & Lönnroth, 1984; Lange *et al.*, 1987; Göransson *et al.*, 1993). In pigs, an amount of only one picomole (10^{12} mole) ASP in the blood causes substantial reduction of cholera toxin- and E.coli LT - induced intestinal secretion (Lange *et al.*, 1987). ASP is synthesized in the central nervous system, accumulating mainly in the pituitary gland from which it is transported via the blood and bile to the gut (Lönnroth & Lange, 1985; Lange & Lönnroth, 1986).

Table 3.1 THE EFFECT OF PLANTAGO POLYSACCHARIDES (PPS) IN FEED WITH OR WITHOUT BAYONOX[R] (B) ON THE FREQUENCY OF PWD (GÖRANSSON *et al.*, 1995).

	Without B (5-6w. weaning)		*With B (4w. weaning)*	
	C	C+PPS	C+B	C+PPS+B
No. of herds	9	9	1	1
No. of litters	90	90	15	11
No. of piglets at weaning	870	864	147	106
L.W. at 8 w.,kg	12	13.1	16	16
PWD,% of litters	65	43	77	54

Table 3.2 THE EFFECT OF PROCESSED SUGAR BEET PULP (SBP) ON THE FREQUENCY OF PWD IN 5 TO 6 WEEKS WEANED PIGLETS (GÖRANSSON *et al.*, 1995).

	C	C+SBP
No. of herds	7	7
No. of litters	71	71
No. of piglets at weaning	678	650
Mortality between 3-8w. of age,%	4.3	4.3
L.W. at 8 w.,kg	14.1	14.2
PWD, % of litters	36	21

ASP is transferred from the sow to the developing foeti via the placenta and is also present in the sow's milk (Sigfridson *et al.*, 1995). ASP is absorbed from the intestine and is found with high activity in the blood of the suckling piglets. Directly after weaning the ASP level in the blood of piglets fed an ordinary creep feed drops, but increases again after weaning (Lange *et al.*, 1993). Piglets suffering from clinical PWD have low levels of ASP in their blood compared to clinical healthy littermates (Lönnroth *et al.* 1988).

FEED INDUCED ASP

The production and release of ASP can be increased by balancing normal feed ingredients with sugar, sugar alcohols and pure amino acids (Lönnroth & Lange, 1986). By using antibodies against entero-toxin induced ASP it is possible to detect the level of feed-induced ASP.

The feed-induced ASP has the same anti-secretory effect as the entero-toxic induced ASP, but differs somewhat in the chemical structure (Lönnroth & Lange,

1988). A series of on-farm experiments with weaners was performed with an ASP - inducing diet. In all units the pigs on the ASP - inducing diet demonstrated an increased total amount of ASP in the blood and a major reduction in the incidence of clinical PWD was registered (Table 3.3, Göransson *et al.*, 1993).

Table 3.3 THE INFLUENCE OF AN ASP-INDUCING DIET IN THE PRODUCTION PERFORMANCE AND FREQUENCY OF PWD IN 5 FARMS (GÖRANSSON *et al.*, 1993).

Farm	No. of pigs	DWG,g 0-35days after weaning		ASP,units/ml plasma 4 d. after weaning		PWD[1],%	
		C	ASP-diet	C	ASP-diet	C	ASP-diet
1	303	-	-	0.30	0.92	60	17
2	169	-	-	0.31	0.91	66	27
3	40	202	247	0.42	0.87	35	10
4	54	266	325	0.79	1.05	15	4
5	325	284	380	0.76	0.94	31	2

1) Farm 1-3, % of litters. Farm 4 and 5, % of piglets individually medical treated.

In a large commercial unit the post weaning mortality was reduced considerably when using an ASP - inducing diet (Göransson *et al.* 1993).

It is also possible to induce ASP production by mixing an appropriate amount of sugars and pure amino acids in the drinking water of piglets. A split litter experiment demonstrated a significant increasing effect on the ASP - blood concentrations by such an addition to the drinking water (Table 3.4, Göransson *et al.*, 1995). The DWG during the first week after weaning in the control group was significantly lower compared to the ASP- experimental groups. These findings strongly indicate subclinical forms of intestinal disorder. The results tentatively suggest that the ASP blood levels are significantly correlated to the production performance of piglets during the first week after weaning, although no clinical signs of diarrhoea were seen.

Bolduan (1997) also demonstrated improved post weaning daily weight gain and reduced frequency of PWD by feeding an ASP-inducing diet or diets with sorbic or fumaric acid (Table 3.5).

Stress has been proven to decrease rapidly plasma and pituitary levels of ASP in rats (Lönnroth *et al.*, 1988). In piglets, plasma ASP decreases directly after weaning which may be directly related to an effect of stress. The plasma feed-induced ASP-concentration, however, does not drop during the first week post weaning, provided that the piglets eat an ASP-inducing diet (Göransson *et al.*, 1993). This difference in weaning stress sensitivity is very important when evaluating the prophylactic effect against PWD of ASP-inducing diets.

Table 3.4 FEED INDUCED ASP-BLOOD CONCENTRATIONS AND PRODUCTION PERFORMANCE FOR PIGLETS WEANED AT 32 DAYS OF AGE (GÖRANSSON *et al.,* 1994).

	C	*ASP Diet type A*	*ASP Diet type B 10d. post w. then C*	*C+ASP-sol 5d prior and 5d post weaning*
No of piglets	28	28	28	28
DWG, g 14-32 days	274	280	273	260
32-39days	92[a]	155[b]	187[b]	162[b]
39-63days	504[ab]	525[ab]	542[a]	498[b]
Daily energy intake, MJ ME				
32-39 days	4.51[a]	5.38[b]	5.67[b]	4.39[a]
39-63 days	10.1	10.7	10.4	10.1
PWD, % of piglets	0	0	0	0
Feed induced ASP, units/ml plasma				
5 days post weaning	0.12	1.32	1.20	1.20

a,b) Means with different letters differs statistically significant (P<0.05).

Table 3.5 THE EFFECT OF DIFFERENT ADDITIVES ON PERFORMANCE DURING 4 WEEKS POST WEANING AND PWD 12% SINGLE-CAGED WEANERS PER TREATMENT WITH AN AVERAGE WEANING WEIGHT OF 9.7 KG (BOLDUAN, 1997).

	Control diet	*Formic acid 0.65%*	*Sorbic acid 1.80%*	*ASP-inducing diet*	*olaquindox 50 ppm*
Feed intake, g/day	1079	1099	1096	1077	1051
DWG, g	426	500	537	511	453
FCR kg/kg	2.62	2.24	2.05	2.14	2.36
PWD, No of animals affected x days					
1st week post weaning	8	2	4	8	3
2nd week post weaning	10	6	0	6	24
3rd week post weaning	21	6	5	6	0
4th week post weaning	0	0	0	0	0

The technology regarding the composition of ASP-inducing diets is patented (Swedish patent no. 9000028-2) and is already used commercially in Swedish pig diets.

ZINC

For many years Zinc in form of ZnO has been used for the prevention of PWD in Denmark. Damgaard Poulsen (1989) reported a significant effect of 2500–4000 mg Zinc as ZnO per kg diet on the severity of PWD as well as on daily weight gain (DWG). In another Danish investigation Holm (1989) fed 2400 ppm ZnO in the diet during 14 or 28 days post weaning. The incidence of PWD was significantly reduced, but the DWG was not affected. No toxic symptoms in the pigs were observed in either of these two investigations. Holmgren (1994) compared 122–176 ppm olaquindox and 1500–2500 Zn as ZnO with a control diet on 6 farms. Zinc was as good a prophylactic agent as olaquindox and no difference in the growth promoting effect could be seen. The frequency of PWD was very high in piglets receiving only the control diet (Table 3.6).

Table 3.6 THE EFFECT OF ZnO OR BAYONOX ON POST WEANING DWG AND FREQUENCY OF PWD (HOLMGREN, 1994).

Farm[1]	Control diet		ZnO-diet		Boyonox-diet	
	DWG, g	PWD %	DWG, g	PWD %	DWG, g	PWD %
A	200	18	220	2	250	2
B	220	32	220	1	240	1
C	400	52	420	10	420	1
D	460	45	520	0	500	0
E	300	80	350	0	350	8
F	280	60	325	12	325	12

Feeding

Invariably *ad libitum* feeding of non-antibiotic diets during the first 7 to 10 days post weaning is difficult to apply in practical production. In an experiment by Rantzer *et al*. (1995) a small restriction in feed allowance during the 3rd to 8th day after weaning at the age of 5 weeks significantly reduced the incidence of diarrhoea (Table 3.7).

In a practical situation the general rule is to feed the weaners with an amount of feed corresponding to 2–3% of the body weight during the first week post weaning. After that the allowance should be increased successively to *ad libitum*. During the restricted period it is essential that feed is available for as many piglets as possible at the same time. Most often piglets are fed on the floor *via* a dispenser operated by the piglets.

Table 3.7 *AD LIBITUM* FEEDING COMPARED TO RESTRICTED FEEDING OF PIGLETS DURING THE 3RD TO 8TH DAY AFTER WEANING, SHOWING THE PERFORMANCE DURING THE FIRST 14 DAYS POST WEANING (RANTZER *et al.,* 1995)

	Ad libitum	Restricted	Level of significance P<0.05=S
No. of litters	16	16	
DWG, g	147	122	S
DFI kg/pig	0.29	0.24	S
Diarrhoea score[1]	0.52	0.35	S
No. of antibiotic treatments/pig /day	0.23	0.15	S
Pigs with dominance of haemolutic *E.coli,* %	50	28	S

[1] 0=no diarrhea and 3=severe diarrhea

Housing and management

It is of utmost importance that the piglets are heavy at birth and well-fed by the sow so that their weight gain until weaning is rapid in order to minimize the risk of diarrhoea. Over the last few years the importance of an adequate feed and feeding program for the sow in order to minimize the risk of hypoagalactia and to optimize milk production has been recognized. *Ad libitum* feeding of lactating sows has proved to increase the weaning weight of the piglets (Table 3.8, Sigfridson, 1992). Age at weaning is closely correlated with the occurrence of PWD (Svensmark *et. al,* 1989) as is the weaning weight of the piglets (Göransson, 1989; Svensmark, 1989).

Table 3.8 THE EFFECT OF AD LIBITUM FEEDING OF SOWS ON LITTER PERFORMANCE (SIGFRIDSON, 1992).

	Restricted feeding	Ad libitum feeding	Level of significance P<0.05=S
No. of litters	77	69	
No. of piglets born	12.10	11.70	NS
weaned	10.90	10.70	NS
Piglet weight at birth, kg	1.42	1.37	NS
at weaning (38 days)	11	11.80	S

Experimentally, Swinkels *et al.* (1988) demonstrated the negative effect of varying air temperature in the pen on the incidence of PWD. Swedish pig farmers have learned the benefits of a warm and dry environment, free of draught as being even more important when antibiotic-free weaning diets are fed.

Since the start of the Swedish ban most pig farmers have organized their flow of sows and piglets in strict "all in all out" systems allowing for cleaning and health care programs to be applied.

The most common housing system includes a farrowing pen from which the sow is moved at weaning. Seven to ten days later the piglets are moved to weaner pens. In the second most common system the piglets are left in the farrowing pen until they reach a live weight of 25–35kg. Locally a multi-suckling system on a deep straw bed has become very popular. In this system the sows and their litters are moved from the farrowing unit to a big pen in groups of 7–15 sows. The sows are fed *ad libitum* and at weaning the sows are moved and the piglets remain on the deep straw bed pen until they reach a weight of 25–35 kg.

Some farmers have invested in a very exclusive system in which the pigs are taken to slaughter weights in the same pen in which they were born. In spite of the high cost of this system it may ultimately recoup the investment due to very high production performance. Holmgren and Lundeheim (1994) found the amount of Bayonox[R] prescribed on PWD diagnosis to be highly correlated to the rearing system.

Genetics

Stigson (1979) reported PWD to have a heritability of 0.17. These results were supported by Göransson (1989) who found a highly significant effect of father and grandfather on the incidence of PWD.

Feeding pigs in practice without antibiotics

The structure of the Swedish pig production industry has changed quite rapidly since 1986. The herd size has been increasing, programed production systems have been implemented and weaning at 28-35 days of age is currently practiced.

Swedish pig producers have learnt to wean their piglets without the use of antibiotics although some medication is still prescribed by veterinarians. Successful weaning without antibiotics requires healthy piglets heavier than 8 kg at weaning, good pig house climate, careful management and a well-balanced diet, usually restricted-fed.

An important experience is that young farmers tend to deal better with the non-antibiotic situation compared to the elder ones who remember the "the good old days with antibiotics" when PWD was no problem.

Weaning diets in Sweden are all of low CP content. Different kinds of fibre are used, mainly beet pulp fibre. Some diets include the ASP concept, and all of them some kind of acid. Several experiments have been performed with different kinds of probiotics and feed enzymes in the Swedish type of diets. These experiments have, however, not demonstrated any positive results useful for practical pig production. Since 1992 the use of 2000 ppm Zn in the form of ZnO for the first 2 weeks after weaning has been allowed temporarily. Many producers feed this kind of diet for the prevention of diarrhoea.

This use of ZnO is the subject of much debate with respect to the accumulation of Zn in the soil, however, and as the permit is given as an exemption from the national situation it is believed to be of short duration.

No investigations concerning the effect of non-antibiotic feeding on the growing/finishing pig performance have been undertaken. According to surveys of pig performance records in Sweden, neither morbidity nor production performance have been affected by excluding the growth promoters from the feed. Traditionally Swedish pigs have been brought to slaughter from 25kg in strictly "all in all out" systems. This factor is believed to be of major importance when explaining why the Swedish growing/finishing pig production performance is fully comparable to the Danish where growth promoters are used, but a continuos flow of pigs through the units still is common. Experimentally the effect of carbadox on pig performance in good and bad health situations has been very clearly demonstrated by Stahly et al. (1994, Table 3.9). According to their trial, health condition has a higher rank of importance than the use of a growth promoter. The growth-promoting effect was much higher in the low health status group of pigs than in the high health status group.

Norwegian and Swedish growing/finishing pig trials with additions of formic acid or salts of formic acid to diets without growth promoters, have consistently shown positive effects on daily gain and feed conversion (NFK-feed brochure; LFU unpublished data). At present, NFK, the largest feed compounder in Norway, add formic acid in commercial feed for growing finishing pigs and sows. All Norwegian pig diets have, since 1995, been produced without growth promoters according to an agreement beween pig producers, feed manufacturers and the slaughter industry.

Table 3.9 RESPONSES OF PIGS WITH A LOW OR HIGH LEVEL OF ANTIGEN EXPOSURE TO DIETARY ANTIMICROBIAL AGENTS[1] (STAHLY *et al.*, 1994).

	Antigen exposure	carbadox, ppm	
		0-0	*50-0*
DWG, g	Low	845	858
	High	686	735
FCR	Low	2.75	2.67
	High	3.12	2.9
Backfat, mm	Low	27.4	26.4
	High	32.2	27.7
L. Muscle area, cm^2	Low	38.4	39.5
	High	32.6	35.8
Carcass muscle, %	Low	56	57.6
	High	51.9	54.2

1) 0 or 55 ppm carbadox from 5 to 34 kg body weight and from there to slaughter at 115 kg live weight all pigs on control feed with no carbadox.

Conclusions

Excluding growth promoters from pig diets in Sweden mainly affected the post weaning weight gain and the frequency of diarrhoea. Although antibiotic medication *via* the weaner diets increased after the ban, the total amount of antibiotics in pig production has been substantially reduced.

The ban has initiated a debate focused on the importance of correctly balanced diets, health status, pig house environment and management. Pig farmers currently pay considerable interest to planning their production according to an "all in all out" system.

Finally "the Swedish example" proves that it is completely possible to produce pig meat efficiently without the use of growth promoters, although increased attention must be paid to feed, feeding, environment and management.

References

Ball. R. O. and Aherne, F. X. (1987). Effect of diet complexity and feed restriction on the incidence and severity of diarrhoea in early weaned pigs. *Canadian Journal of Animal Science*, **62**, 907–913.

Bolduan, G., 1997. Biöffütlerung beim Absetzferkel. *Archiv für Tierzucht A*, **40**, 95–100.

Björneroth, L., Franklin, A. and Tysén, E. (1995). Animal consumption of anti bacterial and anti parasitic drugs in Sweden during 1988 to 1993. Submitted for publication.

Damgaard Poulsen, Hanne. (1989). Zinkoxid til grise i fravaenningsperioden. Meddelelse Nr 746, Statens Husdyrbrugsforsog. Foulum. Denmark.

Danielsen, V. (1984). Effekten av reduseret proteintilldelning till smågrise. *Hyologiosk Tidskrift.*, **12**, 16–19.

Göransson, L. (1989). The effect of very restricted feeding during late pregnancy on the production performance of sows and litters. Thesis, *Swedish University of Agricultural Science Report* 188, Uppsala.

Göransson, L., Martinsson, K., Lange, S. and Lönnroth, I. (1993). Feed-induced Lectines in Piglets. *Journal of Veterinary Medicine,* **B40**, 478–484.

Göransson, L., Lange, S. and Lönnroth, I. (1995). Post weaning diarrhea : Focus on diet. *Pigs News and Information,* **16**, 3, 89N–91N.

Holm, A. (1989). Zinkoxid tilsat foderet til forebyggelse af E.coli-betinget fravaenningsdiarré. Hyologisk Tidskrift 1. Denmark.

Holmgren, N. (1994). Inblandning av zinkoxid i smågrisfoder som profylax mot avvänjningsdiarre. Praktiskt inriktade grisförsök, Nr 1, Skara, Sweden.

Holmgren, N. and Lundeheim, N. (1994). Djurhälsomässiga behovet av fodermedelsantibiotika i smågrisproducerande besättningar. *Sv. Vet. Tidn.*, **46**, 57–65.

Lange, S. and Ek, E. (1995). On putting the argument for banning or tightly controlling the use of antibiotics as feed additives. *Proceedings of Word's Poultry Science Association*, Turkey.

Lange, S., Martinsson, K., Lönnroth, I. And Göransson, L. (1993). Plasma level of ant-secretory factor and it´s relation to post weaning diarrhoea in piglets. *Journal of Veterinary Medicine*, **335**, 113–118.

Lange, S. and Lönnroth,I. (1984). Passive transfer of protection against cholera toxin in rat intestine. *Microbiology letters*, **24**, 165–168.

Lange, S. and Lönnroth, I. (1986). Bile and Milk from cholera toxin tested rats contain a hormone- like factor which inhibits diarrhoea induced by toxin. *Int. Archs Allergy Appl. Immun.*, **79**, 270–275.

Lange, S., Lönnroth, I. and Skadhauge, E. (1987). Effect of anti secretory factor in pigs. *Europ. Physiol. (Pflugers Archs)*, **409**, 328–332.

Larsen, J. L. (1981). Effect of pectin on secretion in pig jejunal loops challenged to enteropathogenic E. coli or enterotoxin (LT). Nord. *Vet.-Med.*,**33**, 218–223

Lönnroth, I. and Lange, S. (1985). A hormone-like protein from the pituitary gland inhibits intestinal hyper secretion induced by cholera toxin. *Regulatory Peptides*, Suppl. **4**, 216–218.

Lönnroth, I. and Lange, S. (1986). Purification and characterisation of anti secretory factor a protein in the control nervous system and in the gut which inhibits intestinal hyper secretion induced by cholera toxin. *Biochemica et Biophysical Acta*, **883**, 138–144.

Lönnroth, I. & Lange, S. (1988). In: Social stress of Pigs. Symposium Abstract, Pharmacia Leo Therpeutics, Malmö, Sweden.

Lönnroth, I., Martinsson, K. and Lange, S. (1988). Evidence of protection against diarrhoea in suckling piglets by a hormone like protein in sow´s milk. *Journal of Veterinary Medicine*, **335**, 628–635.

Miller, B., Newby, T.J., Stokes, P.D, Hampson, D. and Bourne, F.J. (1983). The role of dietary antigen in the aetiology of post weaning diarrhoea. *Annales de Reserches Veterinaires*, **14**, 487–492.

Prohaszka, L. and Baron, F. (1980). The predisposing role of high dietary protein supplies in enteropathogenic E. coli infection of weaned pigs. *Zbl. Vet. Med.*, **27**, 222–232.

Rantzer, D., Svendsen, J. and Weström, B. (1995). Effect of feeding strategies on pig performance and health during the post weaning period. *Proceedings of EAAP*, Prague, Tjeckoslovakia.

Robertsson, J. (1994). Prohibited use of antibiotics as feed additive for growth promotion effects on piglet health an production parameters. *Proceedings of IPVS*, Bangkok, Thailand.

Sigfridsson, K. (1992). The effect of sow feeding regime on sow and litter performance. Internal report. Swedish Pig Center, Svalöv, Sweden.

Sigfridson, K., Lange S. and Lönnroth, I. (1995). Anti Secretory Protein and Feed Induced Lectines in sow and suckling piglet. *Abstracts 46th Annual Meeting of EAAP*, Prague, Tjeckoslovakia.

Stahly, T.S., Williams, N.H., and Zimmerman, D.R. (1994). Impact of carbadox on rate and efficiency of lean tissue accretion in pigs with low or high immune system activation. *Journal of Animal Science*, **72** (Suppl. 1):84.

Stigson, M. (1979). Grisningsbeteende och smågrisdödlighet. Försöksledarmötet. Kons. avd. rapporter, Allmänt 17, Uppsala, Sweden.

Svensmark, B., Nielsen, K., Willeberg, P. and Jorsal S-E. (1989). Epidemiological studies of piglets diarrhoea in intensively managed Danish sow herds. *Acta vet. scand.*, **30**, 35–62.

Swinkels, W.G.M., Kornegay, E.T. and Verstegen, M.W.A. (1988). The effect of reduced nocturnal air temperature and feed additives on the performance, immune response and scouring index of weaning pigs. *Journal of Animal Physiology and Animal Nutrition*, **60**, 137–145.

II

Poultry Nutrition

4

EFFECTS OF DIFFERENT FACTORS INCLUDING ENZYMES ON THE NUTRITIONAL VALUE OF FATS FOR POULTRY

C.W. SCHEELE, C. KWAKERNAAK, J.D. VAN DER KLIS AND G.C.M. BAKKER
Institute for Animal Science and Health (ID-DLO), PO Box 65, 8200 AB, LELYSTAD

Introduction

The continuously increasing production rate of modern broiler and layer strains requires a high daily intake of energy, which can be achieved by feeding high energy diets. The apparent metabolizable energy (AME) content of dietary fats is almost three times as high as that of other feedstuffs. Therefore fat is almost essential in the formulation of those high energy diets. Many types of fats are available for use in poultry diets. The digestibility of fats, and thus the AME, depends on the chemical and physical characteristics of different fat sources (Freeman, 1976; Freeman, 1984; Krogdahl, 1985).

Young chickens in particular have difficulty in digesting high contents of saturated fats in diets (Carew *et al*, 1972; Fedde *et al*, 1960; Wiseman and Salvador, 1989). Hard fats are characterized by high melting points, which can be related to high concentration of long chain saturated fatty acids in the fats. In the experiments of Renner and Hill (1960), it was shown that young chickens have a limited ability to digest and absorb fats with a high percentage of palmitic acid (C 16:0) and stearic acid (C18:0).

The most common monounsaturated fatty acid in dietary fats is oleic acid (C18:1) having one double bound. High contents of polyunsaturated fatty acids, such as linoleic acid (C18:2) with two double bounds and linolenic (C18:3) with three double bounds are present in vegetable oils. Unsaturated fats have lower melting points than saturated fats.

Experiments of Gomez and Polin, (1976) and Kussaibati *et al.* (1982) demonstrated that the addition of bile salts to diets containing high concentrations of saturated fatty acids improved the metabolisable energy values of such diets in young chicks. The poor utilization of saturated fats in chickens can be attributed to an interaction between a high melting point of these fats and a small bile salt pool in the intestinal tract due to a low rate of bile salt production in young chicks.

Fats or triglycerides are hydrolysed in the intestine to monoglycerides, fatty acids and glycerol, which are subsequently absorbed. Fats are insoluble in water but they can be emulsified in the chyme. Saturated fatty acids, such as stearic and palmitic acid, in particular require an emulsifier (bile salts) in the intestinal tract (Garrett and Young, 1975). Furthermore after hydrolysis of fat, fatty acids are water-soluble in the presence of bile. Bile is necessary for a normal absorption of long chain saturated fatty acids from the chyme.

It has long been alleged that the digestibility of saturated fats can be improved by mixing them with liquid oils. Liquid oils with high contents of polyunsaturated fatty acids (PUFA) will mix better with the chyme in the intestinal tract than saturated fats. Thus in the form of an emulsion, PUFA will have fewer impediments to hydrolysis in the intestine than larger particles of saturated fats.

Emulsification improves fat hydrolysis by increasing the surface area of fat droplets to the enzymic action of lipase. Unsaturated oils hydrolysed to monoglycerides, will enhance the emulsification of other fat particles. Monoglycerides act as emulsifiers in the chyme (Freeman, 1984). Therefore, it has been assumed that blends of saturated fats and liquid oils have better digestibility values in chickens than can be calculated from digestibility of the separate components in the mixture. This so-called synergistic effect in blends has been used, albeit with little supporting data, to increase the energy values of animal fats and palm oil, which have high contents of palmitic and stearic acid, by blending with soyabean oil, which has a high content of PUFA.

The digestibility of fats can be influenced by other organic feed components affecting physical characteristics of the chyme. An increased viscosity of the chyme could impede the action of the available small quantities of bile salts and other emulsifiers within the gut. Hesselman and Åman (1986) demonstrated a relationship between intestinal viscosity and absorption of organic feed components. High molecular weight carbohydrate complexes, such as water soluble pentosans (WSP), were shown to increase the viscosity of the fluid phase of the chyme in broilers (Bedford *et al.*, 1991).

A distinct relationship between the ileal viscosity and fat digestibility in broilers was found by van der Klis *et al.* (1995b). Bedford *et al.* (1991) showed that an enzymatic depolymerization of carbohydrate complexes by a pentosanase decreased their viscous nature and improved broiler performance. Van der Klis (1995b) found an improvement in fat digestibility by dietary endoxylanase addition, which was related to a decrease in chyme viscosity.

Different experiments at our institute were conducted to determine factors which affect the digestibility and AME values of fats and oils in poultry diets.

Experimental procedures

Experimental diets were supplied to groups of broilers, which were housed in battery cages (12 birds per cage). Each experimental diet was given to 6 cages. Feed and water were continuously available. Droppings were collected once a day for four consecutive days. Gross energy, dry matter, nitrogen and fat content of feed and excreta were analysed in order to calculate AME and fat digestibility. AME values of fats were calculated either from digestibility values or by subtracting the AME value of a basal diet without added fat from the AME value of the same basal diet with added fat. Chemical and physical characteristics of feedstuffs , diets, chyme and excreta were determined according van der Klis (1995a). Experiments with adult cocks and with laying hens were carried out with birds kept individually in battery cages. Similar procedures were followed as used in the experiments with broiler chickens.

AME values of fats and oils in poultry

Maize - soyabean basal diets without or with 90g added animal fat/kg were fed to broilers from 0 - 8 weeks of age and to adult cocks. AME values of diets and digestibility values of the fats were determined at 2, 4, 6 and 8 weeks of age in broilers and once in adult cocks. Approximately 30% of the total fatty acids in the supplemental fat was palmitic acid + stearic acid. The determined AME values of the basal diet and of added fat and fat digestibility values are given in table 4.1.

Table 4.1 shows no significant effect of the age of broilers on the AME value of the basal diet. However an important significant effect of age on AME and digestibility of fat was found. These findings are in agreement with the results of Renner and Hill (1961) which concluded that young chickens especially have a limited ability to digest fats with high contents of palmitic and stearic acid, and Carew *et al.* (1972), Fedde *et al.* (1960), Wiseman and Salvador (1989) on the effect of age.

In an experiment with broilers of 4 weeks of age AME values of different oils and fat were determined. The results given in table 4.2 were related to the percentages of PA + SA (palmitic and stearic acid) and of PUFA (polyunsaturated acids) in the fats.

Table 4.1 AME VALUES OF A BASAL DIET AND OF AN ADDED ANIMAL FAT AND DIGESTIBILITY VALUES OF DIETARY FAT IN BROILERS AT DIFFERENT AGES AND IN ADULT COCKS

	AME		Total fat digestibility %
Age	*(1) Basal diet* MJ/kg	*(2) Fat (added)* MJ/kg	
Week 2	10.76[a1]	27.96[a]	62.5[a]
Week 4	10.94[a]	29.02[b]	66.9[b]
Week 6	10.92[a]	32.40[c]	72.0[c]
Week 8	11.00[a]	33.19[d]	73.4[c]
Adult cocks	11.70	36.47	85.0

1) within the broiler experiment: different superscripts within the same column denote significant differences (P< 0.05)

Table 4.2 AME VALUES OF FATS WITH DIFFERENT CONTENTS OF PUFA (POLYUNSATURATED FATTY ACIDS) AND PA + SA (PALMITIC + STEARIC ACID) DETERMINED IN 4 WEEK OLD BROILERS.

Oils and fats	*AME* MJ/kg	*PUFA* %	*PA+SA* %
Soyabean oil	35.4[a1]	60.0	15.3
Safflowerseed oil	35.0[a]	76.2	10.0
Grapeseed oil	35.0[a]	69.9	9.3
Linseed oil	34.0[ab]	74.7	8.4
Rapeseed oil	33.5[abc]	32.7	6.9
Olive oil	32.5[bc]	18.3	15.8
Coconut oil	31.6[bc]	10.2	4.0
Groundnut oil	31.5[cd]	33.5	15.7
Poultry fat	30.1[d]	16.4	23.3
Mixed animal fat	28.1[e]	9.3	31.6
Palm fat	25.8[f]	11.0	45.6
Tallow	24.5[f]	7.9	39.9

1) different superscripts within the same column denote significant differences (P<0.05)

By means of regression analyses, relationships between the fatty acid composition of the oils and fats and the AME values were calculated. The calculated relationships are represented by two equations.

AME (MJ/kg) = 36.4 - 0.26 (PA +SA)   (1)
$r^2 = 0.85$ RSD = 1.5 MJ

AME (MJ/kg) = 28.0 + 0.10 (PUFA) MJ/kg   (2)
$r^2 = 0.60$ RSD = 2.4 MJ

The best results were obtained by means of the equation predicting the AME from the palmitic + stearic acid content (equation 1). Rapeseed oil and olive oil with relative low values for PUFA have nevertheless high AME values. This observation is related to high values of monounsaturated fatty acids (predominantly oleic acid) in these oils. Coconut fat also with a low PUFA content, has a high AME value because of the high contents of short chain fatty acids that are well digested. These data are all in agreement with the observations of Wiseman and Blanch (1994).

Synergistic effects between oils and fats fed to poultry in a mixture

It has been demonstrated by several research workers that the AME of a lipid can be altered by feeding it in a mixture with other lipids. Sibbald (1978) found that by adding soyabean oil to tallow, the ME values of the mixtures were higher than the sum of the ME values of its components parts. Data of Lewis and Payne (1966) revealed a curvilinear increase in ME values of tallow- soyabean oil mixtures as the level of soyabean oil in the mixture increased linearly from 0 to 30%.

Mixtures of animal fats, or so called renderers fats, are available in large quantities to the animal feed industry. The world annual production of tallows is about 6 million tons. In the United States, more than 50% of all tallows and greases produced are now consumed by domestic feed producers (National Renderers Association). As the price of these fats is generally considerably lower than of vegetable oils, renderers fats provide an economical source of dietary energy. Moreover some vegetable fats have become economically attractive during the last decade. In several countries of Latin America and in South East Asia, large quantities of palm fats are available now at relatively low prices to the world animal feed industry. Current annual production of crude palm oil in a small country like Costa Rica in Central America is approximately 80.000 tons (Scheele *et al.* 1995).

However, both renderers fat and palm oil are saturated, having relatively low AME and digestibility values for poultry as is shown in table 4.2. By blending these saturated fats with oils having low melting points it has been suggested that the energy value of the low cost fats can be improved. This phenomenon, whereby the dietary energy value of a saturated fat may be improved through blending with a more unsaturated fat, is referred to as synergism. However, detailed investigations by Wiseman and Lessire (1987) and Wiseman and Salvador (1991) failed to confirm such a response and Wiseman (1990) argued that synergism between fats (as opposed to fatty acids) was a conceptually unsound principle. However, investigations into it are still proceeding. This synergistic effect of adding different kinds of oils to animal fat mixtures (renderers fat) was studied in 3 week old broiler chickens (Scheele and Versteegh, 1987). The experimental diets consisted of 90% basal diet and 10% supplemental fat. The experimental results are shown in table 4.3.

Table 4.3 AME VALUES OF RENDERERS FAT (RF), PURE OILS, AND OF BLENDS OF 70% RF AND 30% OF AN OIL

RF and oils	Determined AME single components MJ/kg	blends 30% oil + RF MJ/kg	AME of blends calculated from component parts MJ/kg	Synergistic improvement %
RF	22.5	-	-	-
Soyabean oil	32.8	29.9	25.6	16.8
Safflower oil	32.4	30.0	25.5	17.6
Linseed oil	31.5	28.9	25.2	14.7
Rapeseed oil	31.1	26.7	25.1	6.4
Olive oil	30.5	23.0	24.9	- 7.6
Coconut oil	29.3	27.6	24.5	12.6

The values given in table 4.3 show that oils with high contents of PUFA had important synergistic effects on the AME of the renderers fat. In addition coconut oil with about 70% saturated short chain fatty acids ($C8:0 + C10:0 + C12:0 + C14:0$), appeared to have a positive effect on the AME of RF.

Rapeseed and olive oils are characterized by high contents (more than 50%) of monosaturated oleic acid. The difference between these fats is that the PUFA of rapeseed oil is nearly twice as high as found in olive oil. Oleic acid seems to have no synergistic effect on the AME of RF. The fatty acid composition of olive oil obviously did not complement that of RF, as the synergistic improvement of olive oil on RF was found to be negative. If it is assumed that the AME of a blend cannot be higher than the AME of the component with the highest value, this gives another means of comparison of the synergistic effects within blends. Maximum synergistic effects of soyabean oil and coconut oil would be respectively $(32.8 - 25.6)/25.6 = 0.281 = 28.1\%$ and $(29.3 - 24.5)/24.5 = 0.196 = 19.6\%$. Thus the synergistic effect of 16.8% of soyabean oil represents 60% of the maximum value of 28.1%. The synergistic effect of 12.6% of coconut oil represents 64% of the maximum value of 19.6%. From these calculations it can be concluded that the synergistic improvement of short chain fatty acids in these blends is at least as important as the synergistic effect of PUFA.

Other fat sources available for the animal feed industry are byproducts of the oil industry. Several byproducts are obtained during the process of extraction and refining in the palm fat and soyabean oil industries. An important component of these byproducts are free fatty acids (FFA) which can be used in animal feeds. Studies of the influence of FFA on AME of fats have been undertaken by Young (1961), Renner and Hill (1961) and Wiseman and Salvador (1991) all of which concluded that higher levels of FFA are associated with lower AME values.

Scheele *et al.* (1995) and Zumbado *et al.* (1996) studied the use of free fatty acids in poultry diets. The studies were financed by the ISC programme of the European Commission in Brussels.

Palm free fatty acids (PFFA) can be considered as a saturated fat with low AME value for poultry. A study was carried out to find out synergistic effects of soyabean free fatty acids (SBFFA) in a mixture with PFFA. In this way, a low cost vegetable fat blend having a high AME value could be formulated for the poultry feed industry.

AME values were determined in 4 weeks old broiler chickens. The AME values of fats and blends added as 5% to a basal diet are given in table 4.4.

Table 4.4 AME VALUES OF SOYABEAN FREE FATTY ACIDS (SBFFA), PALM FREE FATTY ACIDS (PFFA), AND TWO BLENDS.

Blend 1 = 50% PFFA + 50% SBFFA; Blend 2 = 75% PFFA + 25% SBFFA
All fats were added as 5% to a basal diet

Fats and blends	Determined AME values	Calculated AME values of blends from component parts	Synergistic improvements
	MJ/kg	MJ/kg	%
PFFA	26.32	-	-
SBFFA	30.02	-	-
Blend 1	29.35	28.17	4.2%
Blend 2	28.77	27.25	5.4%

The results in table 4.4 suggest that AME values of saturated palm oil, particularly palm free fatty acids were not improved in the same way by PUFA from soyabean free fatty acids as was shown in table 4.3 with regard to the improvement of the AME of renderers fat by soyabean oil. However, if the maximum synergistic effects that are possible in these blends are calculated by assuming that the AME of a blend cannot be higher than the AME of its single components, the following results are apparent. Maximum synergistic effects of SBFFA in blend 1 and blend 2 would be respectively (30.02 - 28.17)/28.17 = 6.6% and (30.02 - 27.29)/27.9 = 10.0%. The synergistic improvements of blend 1 and blend 2 given table 4 represent respectively 64% and 54% of the maximum values. From these results it can be concluded that AME values of palm fatty acids can be improved in the same way as was shown in table 3 with respect to renderers fat.

As the AME of soyabean oil is higher than that of SBFFA, higher AME values could be obtained by using soyabean oil in a blend with PFFA.

Nevertheless table 4.4 shows a synergistic improvement of the AME of blends containing palm free fatty acids. Thus byproducts of the palm oil industry could be used in a better way in the feed industry by mixing with other oils.

Interaction between fats and wheat cultivars in poultry diets

It is possible that the feeding values of fat blends with high contents of saturated fatty acids, such as palm oil and renderers fats, could also be negatively affected by other components in diets such as carbohydrate complexes which increase the viscosity of the fluid phase of the chyme in chickens. Therefore attention has to be paid to interactions between fats and grains in diets such as wheat.

It is generally accepted that the AME values of wheat containing diets can be highly variable in broiler chickens. (Mollah *et al.*, 1983; Rôgel *et al.*, 1987; Scheele *et al.*, 1993, 1994; van der Klis *et al.*, 1995b). Choct and Annison (1992) demonstrated that isolated water soluble pentosans (WSP) from wheat, increased the viscosity of the fluid phase of the chyme, which adversely affected the dietary AME value. Table 4.1 demonstrated that young chickens have a limited ability to digest fats with high contents of saturated fatty acids. Therefore, young birds also might be vulnerable to dietary interactions between saturated fatty acids and carbohydrate complexes in wheats, rye and barley increased the viscosity of the chyme. Fengler and Marquardt (1988) found adverse effects of water soluble pentosans from rye, which increase viscosity values of the chyme, on fat digestibility in young chickens. Adding isolated rye WSP to a wheat-based diet reduced the fat digestibility by 20%. These low fat digestibilities were also observed in a rye diet. Interactions between WSP and other components in wheat, barley and rye, which increase the viscosity of the chyme, and different sources of fats in poultry diets, can have significant effects on performances of broilers and layers particularly at high dietary inclusion levels of both cereals and fats. This may have profound adverse effects on growth, egg production and feed conversion ratio in poultry.

In experiments with broiler chickens raised to 3 weeks of age, AME values of a blend of renderers (RF) and soyabean oil (SBO) added to different basal diets without and with different wheat cultivars were determined. Four wheat cultivars (W_1, W_2, W_3, W_4) were selected based on their *in vitro* viscosity of the supernatant of the wheat samples (van der Klis *et al.*, 1995a). Each cultivar was added to a corn - soyabean mixture including vitamins and minerals (BD), resulting in four basal diets containing 50% wheat and 50% of BD: making diets BW_1, BW_2, BW_3 and BW_4.

The basal diet without wheat (BD) and the basal diets with wheat (BW_1, BW_2, BW_3 and BW_4) were supplemented with either 7% RF or with 7% SBO making 10 different fat-containing experimental diets, (BD-RF, BD-SBO, BW_1-RF, BW_1-SBO, BW_2-RF, BW_2-SBO, BW_3-RF, BW_3-SBO, BW_4 and BW_4-SBO). Supernatant viscosity values of wheats and chyme in the jejunum and ileum in birds fed the different wheat-containing diets were determined. The results of the experiments are given in table 4.5.

Table 4.5 AME VALUES OF RENDERERS FAT (RF) AND SOYABEAN OIL (SBO) IN:
1. A basal diet without wheat (BD).
2. Four basal diets (BW_1, BW_2, BW_3 and BW_4) containing 50% of the wheat varieties W_1, W_2, W_3 and W_4 successively
Viscosity values in supernatant of wheats and of chyme in jejunum and ileum of birds fed fat containing diets.

Diets	AME of RF and SBO	Viscosity supernatants		
		wheat	chyme	
			jejunum	ileum
	MJ/kg	m Pas	m Pas	m Pas
BD-RF	29.03[a1]	-	-	-
BD-SBO	34.60[b]	-	-	-
BW_1-RF	21.78[c]	1.38	2.49[a]	3.42[a]
BW_2RF	20.37[d]	1.84	2.94[a]	4.85[b]
BW_3-RF	17.66[e]	1.48	2.62[a]	4.00[ab]
BW_4-RF	18.14[e]	2.16	2.88[a]	4.16[ab]
BW_1-SBO	31.98[f]	1.38	2.61[a]	3.58[a]
BW_2-SBO	31.02[f]	1.84	3.08[a]	5.07[c]
BW_3-SBO	29.16[a]	1.48	3.09[a]	5.02[c]
BW_4-SBO	30.03[g]	2.16	3.32[a]	5.19[c]

1) Different superscripts within the same column denote significant differences (P<0.05)

The AME values of RF and SBO in diets without wheat were reasonably high in young chickens. A profound effect of wheat cultivars on the AME of RF was found. Although the effects differed between the wheat cultivars all AME values of RF in wheat diets were extremely low. Feeding these wheats together with high levels of RF will have a profound effect on the performances of these birds.

Wheats also affected AME values of SBO negatively, but the effect was much smaller than compared with the effect on RF. Viscosities of chyme were not measured in birds fed diets without wheat, but it can be assumed that wheat will have increased the viscosities of the chyme. These high viscosities may have affected the AME values of fats. However, the differences in AME values of RF between diets with different wheat cultivars could not be attributed to the differences in viscosity values of supernatants. The same is valid for differences in AME values of SBO between diets. Obviously, there are other factors in wheat that have an effect on the AME values of fats and oil besides components that increase the viscosity.

The effect of enzymes on AME values of fats in wheat containing diets for broilers

Bedford *et al.* (1991) showed that enzymatic depolymerization of high molecular weight carbohydrate complexes in the gut, decreased their viscous nature and improved broiler performance. Therefore these enzymes could ameliorate the negative effects of wheats on the AME values of fats. Effects of endoxylanase addition to wheat-containing diets on digestibility values of fats were studied in four week old broiler chickens. AME values of fats from feedstuffs (mainly animal fat and maize oil) in diets containing 50% wheat and approximately 4% fat were calculated from determined fat digestibility values. Four different diets were composed by using four different wheat cultivars. The diets were fed without or with 40 ppm Lyxasan® (an endoxylanase). The experimental results are shown in table 4.6.

Table 4.6 AME VALUES OF FATS IN BROILER CHICKENS. EFFECTS OF ENDOXYLANASE IN FOUR DIETS CONTAINING 50% OF DIFFERENT WHEATS (W_1, W_2, W_3 AND W_4) ON THE AME OF DIETARY FAT AND ON ILEAL VISCOSITY (IL.V)

Diet	Endoxylanase			
	AME fat (MJ/kg)		Ileal viscosity (m Pas)	
	0 ppm	*40 ppm*	*0 ppm*	*40 ppm*
W_1	31.01[a1]	33.33[b]	4.4[a]	3.6[bd]
W_2	31.07[a]	32.97[b]	4.2[ab]	3.2[d]
W_3	29.39[c]	33.69[b]	4.9[a]	3.3[d]
W_4	28.95[c]	32.97[b]	5.8[c]	3.1[d]

1) Different superscript for the same characteristic within the same row and in the same column denote significant differences (P<0.05)

The AME values of small amounts of dietary fats in wheat containing diets were not low, but the addition of endoxylanase to the diets improved the AME values of the fats. Endoxylanase addition also reduced the ileal viscosity. The degree to which endoxylanase increased AME values of fats was related to different wheat cultivars such that AME of fats all approached the same value. Table 4.6 also shows that there is a good relationship between ileal viscosities and AME values of fat, although such a correlation was not found in table 4.5.

Another experiment was carried out with five diets using five different wheat cultivars (50% wheat in the diet). In this experiment, the dietary fat content was 5.5%. A blend of equal parts of soyabean oil and animal fat was added to the diets. The diets were fed to broilers without or with 40 ppm endoxylanase (Lyxasan®). AME values of fats and intestinal viscosities were determined in broiler chickens at 4 weeks of age. The experimental results are presented in table 4.7.

Table 4.7 AME VALUES OF FATS IN BROILER CHICKENS. EFFECTS OF ENDOXYLANASE IN FIVE DIETS CONTAINING 50% OF DIFFERENT WHEATS ((W_5, W_6, W_7, W_8 AND W_9) ON THE AME OF DIETARY FAT AND ON ILEAL VISCOSITY (IL.V)

Diet	Endoxylanase			
	AME fat (MJ/kg)		Ileal viscosity (m Pas)	
	0 ppm	40 ppm	0 ppm	40 ppm
W_5	31.06[a1]	32.97[b]	4.13[a]	3.18[a]
W_6	30.90[c]	33.72[b]	5.04[ab]	3.42[a]
W_7	30.74[c]	33.05[b]	4.14[a]	4.58[a]
W_8	28.87[d]	33.00[b]	6.03[b]	3.23[a]
W_9	24.89[e]	29.51[f]	4.28[a]	3.68[a]

1) Different superscript for the same characteristic within the same row and in the same column denote significant differences ($P<0.05$).

The results shown in table 4.7 reveal a distinct positive effect of endoxylanase addition on the AME of added fats in wheat- containing diets. In the first four diets (W_5 - W_8) with decreasing fat AME values, the addition of endoxylanase increased AME values of fat to almost the same high level. A low AME value of fat was found in W_9. Although the PUFA content of the added fat was not low (half of the added fat was soyabean oil), the interaction between the fat and the wheat variety W_9 was obvious and negative.

Ileal viscosity values were decreased by endoxylase addition but only in treatment DW_8 was this effect significant. It was also observed that the increased fat AME in treatment DW_7 by endoxylanase could not be related to a decrease in ileal viscosity. Furthermore, a notably low fat AME value in DW_9 was not consistent with higher ileal viscosities compared with the other dietary treatments.

Besides the effect of viscosity, there is obviously another factor in wheat that can influence the AME of dietary fats. More research is needed to find out which factors in wheat must be eliminated, possibly by other enzymes, to improve the AME values of fats as is shown in treatment DW_9. Other feed enzymes such as lipases might be important for a further improvement of fat digestibility in wheat-containing diets. Proteases could be beneficial by reducing the visco-elastic behaviour of the gluten fraction of wheat protein in the chyme of chickens. Visco-elasticity could interfere with the mixing of ingesta with the digestive secretions.

A third experiment with broilers was carried out using six different wheat varieties. In this experiment, basal diets were composed with 50% from the different wheat cultivars. To these basal diets was added 7% renderers fat (RF). The diets were fed without or with 40 ppm endoxylanase (Lyxasan®). Together with these wheat-containing diets; a diet without wheat (with maize) containing 7% (RF) was also fed to chickens in the same experiment.

The results shown in table 4.8 indicate that AME values of RF were reduced by wheat in diets. Animal fats, such as RF, contain high amounts of saturated fatty acids. Obviously, the saturated fatty acids are not well digested in diets containing a high level of wheat. In the maize diet, the AME of RF was modest and normal for young chickens. The ileal viscosities found in wheat diet treatments were noticeably higher than in the treatment with the maize diet.

Table 4.8 AME VALUES OF RENDERERS FAT (RF) IN DIETS WITH OR WITHOUT WHEAT, DETERMINED IN CHICKENS AT 4 WEEKS OF AGE.
EFFECTS OF ENDOXYLANASE AND SIX DIFFERENT WHEAT CULTIVARS ON THE AME OF RF AND ON ILEAL VISCOSITY (IL.V). SEVEN DIETS WERE GIVEN WITH 7% RF; DM WITHOUT WHEAT, AND W_1, W_2, W_3, W_4, W_5, W_6 ALL WITH 46,5% WHEAT.

| Diet | Endoxylanase | | | |
| | AME (RF) MJ/kg | | Ileal viscosity m Pas | |
	0 ppm	40 ppm	0 ppm	40 ppm
M	28.70[a]	-	1.85[d]	-
W_1	24.93[b]	26.84[e]	4.36[a]	
W_2	24.66[b]	27.32[c]	3.95[a]	3.06[a]
W_3	24.14[b]	26.88[c]	4.02[a]	2.58[c]
W_4	22.95[d]	26.84[c]	5.76[b]	2.70[c]
W_5	21.79[e]	25.81[f]	5.42[b]	3.06[c]
W_6	21.55[e]	28.23[g]	5.94[b]	2.63[c]

1) Different superscript for the same characteristic within the same row and in the same column denote significant differences (P<0.05).

Different wheat cultivars had different effects on the AME values. In this experiment there was a good relationship between decreasing AME values of RF and increasing ileal viscosities. Endoxylanase increased all AME values of fats in wheat diets and decreased ileal viscosities. The effect was similar for all diets except for W_5 and W_6. In W_5, the improvement of the AME of RF by endoxylanase was relatively small, but in W6 with the lowest fat AME the improvement was the highest.

The results indicate that high levels of fats with high contents of palmitic and stearic acid in wheat-containing diets will have low overall fat digestibility values associated with poor fat AME values. These findings are in agreement with results shown in table 4.4. Table 4.8 shows that endoxylanase can improve nutritional values of diets based on fats with high levels of saturated fats but AME values of RF remain at a low level, except in a combination with one particular wheat cultivar (W_6). Thus one must consider the fatty acid patterns of dietary fats especially when added to wheat-containing diets. A combination of a high quality fat and a high quality wheat together with enzymes can create diets with a high energy value necessary to support a high performance in chickens. Conversely high dietary levels of saturated fatty acids (table 4.5) mixed into in poultry diets together with different wheat cultivars could have significant negative effects on performance of broiler chickens.

More research is needed to find out how low cost fats (mostly saturated fats) can be blended with other fat sources to formulate better fatty acids patterns in wheat-based diets for chickens and which enzymes should be used.

Enzymes affecting AME values of fats in diets for laying hens

Table 4.1 illustrated that higher AME values were obtained in adult cocks than in broiler chickens at different ages. Laying hens also tend to have fewer problems with digesting fats, including those based on saturated fatty acids, than those observed in young chickens. However it might be expected that wheats which increase the gut viscosity of chickens will have the same effect in the gut of laying hens. Therefore, a combination of unfavourable factors, such as the fatty acid pattern in the fat together with components in cereals increasing the viscosity of the fluid phase of the chyme, may reduce the digestibility and absorption of fats in laying hens. As a result endoxylanases could also improve the feeding values of dietary fats in laying hens.

The effects of endoxylanase addition to wheat containing diets on the digestibility of fats were studied in two experiments with laying hens. AME values of dietary fats were calculated from determined fat digestibility values.

In experiment 1, two different diets were forumulated using two different wheat cultivars. The diets contained 50% wheat and 5% fat from soyabean oil. The diets were fed to laying hens without or with 500 mg endoxylanase (Bio-feed Plus CT®) per kg of diet. Fat digestibility and gut viscosity were measured at 36 weeks of age.

The results in table 4.9 reveal that endoxylanase resulted in significant positive effects on the AME value of dietary fat in wheat-containing diets for laying hens. In this experiment soyabean oil was used in diets, which is normally very well digested by laying hens. The experimental results indicate that wheat must have reduced the AME value of the soyabean oil in laying hens. Endoxylanase, affecting the carbohydrate components of wheat, was able to increase AME of fat significantly. The results also show that in this experiment the ileal viscosity was highly correlated with the digestibility and thus with the AME value of fat.

Table 4.9 AME VALUES OF VEGETABLE FATS IN LAYING HENS. EFFECTS OF ENDOXYLANASE IN TWO DIETS CONTAINING 50% OF DIFFERENT WHEAT VARIETIES (W_1 AND W_2) ON THE AME AND ON ILEAL VISCOSITY (IL.V.)

| Diet | Endoxylanase | | | |
| | AME fat (MJ/kg) | | Ileal viscosity (m Pas) | |
	0 mg/kg	500 mg/kg	0 mg/kg	500 mg/kg
DWV$_1$	34.96[a1]	35.83[c]	5.7[a]	3.5[b]
DWV$_2$	34.45[b]	35.87[c]	6.5[c]	3.5[b]

1) Different superscript for the same characteristic within the same row and in the same column denote significant differences (P<0.05).

In a second experiment with laying hens another fat source was chosen and added to the diets in a higher level. Again two different diets were formulated using two different batches (B_1 and B_2) of wheat from the same cultivar. The diets contained 50% wheat and 7% of RF which contained a high level of saturated fatty acids. The diets were fed to laying hens without or with 100 ppm endoxylanase (Natugrain®). Fat digestibility and gut viscosity were measured at 36 weeks of age.

The results in table 4.10 show that AME values of RF added at a high level in diets for laying hens are not much lower than the AME values for soyabean oil found in table 8. These results confirm that the effect of differences in fatty acid patterns is small in laying hens in contrast to young chickens. Table 4.10 also indicates that wheat can reduce AME values of fats; this can be altered positively by endoxylanase.

Table 4.10 AME VALUES OF RENDERERS FAT (RF) IN LAYING HENS. EFFECTS OF ENDOXYLANASE IN TWO DIETS CONTAINING 50% OF DIFFERENT WHEATS (W_1 AND W_2) ON THE AME OF DIETARY FAT AND ON THE ILEAL VISCOSITY (IL.V.)

Diet	*Endoxylanase*			
	AME fat (MJ/kg		*Ileal viscosity (m Pas)*	
	0 ppm	*100 ppm*	*0 ppm*	*100 ppm*
DB$_1$	33.84[a1]	34.63[b]	3.5[ac]	2.9[b]
DB$_2$	33.88[a]	34.83[b]	3.7[c]	3.1[ab]

1) Different superscript for the same characteristic within the same row and in the same column denote significant differences (P<0.05).

In conclusion, attention should be paid to the optimisation of fatty acid profiles of fats in formulations containing a high level of cereal grains, such as wheat. The dietary addition of enzymes specific for soluble polysaccharides in cereals can improve the AME of diets and the performance of poultry.

References

Bedford, M.R., H.L. Classen and G.L. Campbell, 1991. The effect of pelleting, salt and pentosanase on the viscosity of intestinal contents and the performance of broilers fed rye. *Poultry Science*, **70**:1571–1577.

Carew, L.B., Machemer, R.H., Sharp, R.W. and Foss, D.C. 1972. Fat absorption by the very young chick. *Poultry Science*, **52**:738.

Choct, M. and G. Annison, 1992. The inhibition of nutrient digestion by wheat pentosans. *British Journal of Nutrition*, **67**:123–132.

Fedde, M.R., Waibel, P.E. and Burger, R.E. 1960. Factors affecting the absorbability of certain dietary fats in the chick. *Journal of Nutrition*, **70**:447.

Fengler, A.I. and R.R. Marquardt, 1988. Water-soluble pentosans from rye:II. Effects on rate of dialysis and on the retention of nutrients in the chick. *Cereal Chemistry*, **65**:298–302.

Freeman, C.P. 1976. Digestion and absorption of fat. *Digestion and Absorption in the Fowl* (K.N. Boorman andB.M. Freeman, eds.), British Poultry Science, Edinburgh, 1976, p. 117.

Freeman, C.P. 1984. The digestion, absorption and transport of fat - Non-ruminants. *Fats in Animal Nutrition* (J.Wiseman, ed.), Butterworths, London, 1984, p. 105.

Garrett, R.L. and R.J. Young, 1975. Effect of micelle formation on the absorption of neutral fat and fatty acids by the chicken. *Journal of Nutrition*, **105**:827–838.

Gomez, M.X. and D. Polin, 1976. The use of bile salts to improve absorption of tallow in chicks, one to three weeks of age. *Poultry Science*, **55**:2189–2195.

Hesselman, K and P. Åman, 1986. The effect of ß-glucanase on the utilization of starch and nitrogen by broiler chickens fed on barley of low - or high viscosity. *Animal Feed Science and Technology*, **15**: 83–93.

Krogdahl, A. 1985. Digestion and absorption of lipids in poultry. *Journal of Nutrition*, **115**: 675 (1985).

Kussaibati, R., J. Guillaume and B. Leclerq, 1982. The effects of age, dietary fat and bile salts and feeding rate on apparent and true metabolizable energy values in chickens. *British Poultry Science*, **23**: 292–403.

Lewis, D. and C.G. Payne, 1966. Fats and amino acids in broiler rations. 6. Synergistic relationship in fatty acid utilization. *British Poultry Science*, **7**: 209–218.

Mollah, Y., W.L. Bryden, I.R. Wallis, D. Balnave and E.F. Annison, 1983. Studies on low metabolizable energy wheats for poultry using conventional and rapid assay procedures and effect of processing. *British Poultry Science*, **24**: 81–89.

Renner, R. and F.W. Hill, 1960. Utilization of corn oil, lard and tallow by chickens of various ages. *Poultry Science*, **39**:849–854.

Rogel, A.M., E.F. Annison, W.L. Bryden and D. Balnave, 1987. The digestion of wheat starch in broiler chickens. *Australian Journal of Agricultural Research*, **38**: 639–649.

Scheele, C.W., C. Kwakernaak, R.J. Hamer and H.J. van Lonkhuijsen, 1993. Differences in wheat AME and protein, fat and carbohydrate digestibilities of wheat and wheat by-products. *Spelderholt Report 617* (in Dutch). Spelderholt, Beekbergen, The Netherlands.

Scheele, C.W., C. Kwakernaak, J.D. van der Klis, 1994. Factors affecting the feeding value of wheat in poultry diets. In: Wheat and wheat by-products realising this potential in monogastric nutrition. Seminar papers Finnfeeds International Ltd, Utrecht, The Netherlands.

Scheele, C.W., C. Kwakernaak, H.J. van Lonkhuijsen and R. Orsel, 1995. The nutrient digestibility in wheat-based diets and physico-chemical chyme conditions in broilers, related to pelletising methods. *Spelderholt report* (in Dutch). Spelderholt, Beekbergen, The Netherlands.

Scheele, C.W., C. Kwakernaak and M.E. Zumbado, 1995. Studies on the use of palm fats and mixtures of fats and oils in poultry nutrition. Part 1. *Survey and analysis of fats and oils and determination of basic nutritional values of pure fats and oils*. ID-DLO, Lelystad. The Netherlands.

Sibbald, J.R., 1978. The true metabolizable energy values of mixtures of tallow with either soyabean oil or lard. *Poultry Science*, **57**:473–477.

Van der Klis, C. Kwakernaak and W. de Wit, (1995a). Effects of endoxylanase addition to wheat-based broiler diets on physico-chemical chyme conditions and mineral absorption in broilers. *Animal Feed Science and Technology* **51**: 15–27.

Van der Klis, C.W. Scheele and C. Kwakernaak (1995b). Wheat characteristics related to its feeding value and to the response of enzymes. *Proceedings of the 10th European Symposium on Poultry Nutrition.* Antalya, Turkey.

Wiseman, J. and Lessire, M. 1987. Interactions between fats of differing chemical content. 1. Apparent metabolisable energy values and apparent fat availability. *British Poultry Science*, **28**, 663–676.

Wiseman, J. and Salvador, F.S. 1989. The influence of age, chemical composition and rate of inclusion on the apparent metabolisable energy of fats fed to broiler chicks. *British Poultry Science,* **30**:653.

Wiseman, J. 1990. Variability in the nutritive value of fats for non-ruminants. In "Feedstuff Evaluation". pp215–234. *Proceedings of the 50th Easter School in Agricultural Sciences*. Edited by J. Wiseman and D.J.A. Cole, Butterworths, London.

Wiseman. J. and Salvador, F. 1991. The influence of free fatty acid content and degree of saturation on the apparent metabolisable energy value of fats fed to broilers. Poultry Science. *Poultry Science*, **70**, 573–582.

Young, R.J. 1961. The energy value of fats and fatty acids for chicks. *Poultry Science*, **40**: 1225–1233.

Zumbado, M.E., C.W. Scheele and C. Kwakernaak, 1996. Studies on the use of palm fats and mixtures of fats and oils in poultry nutrition. Part 2. *Digestibility studies and evaluation of broiler and laying hen performance* CINA Zootecnia. University of Costa Rica, San Pedro, Costa Rica. ID-DLO Lelystad, The Netherlands.

5

FEEDING THE MALE TURKEY

M.S. LILBURN
Animal Sciences Department, The Ohio State University, Wooster, OH 44691, USA

Introduction

The success that the turkey industry has achieved over the last decade has been largely due to continual improvements in growth and carcass yield and innovative ways of making turkey meat into a wide array of further processed products. The latter factor has coincided with increased perception of poultry as both a highly nutritious and low fat product. The breast portion is the most important part of the carcass from a further processing standpoint and it is primarily made up of two muscles, the *Pectoralis major* and *Pectoralis minor*. The *P. Major* is that portion of the breast which has been most responsive to selection for increased breast yield (Lilburn and Nestor, 1991). At a typical market weight, these two muscles account for approximately 18% and 4.5% of live weight, respectively, and they are easily dissected from the carcass for quantification of breast component yields. This approach to breast yield determination between research stations is encouraged because it allows for easier comparisons of experimental data generated at different locations.

Within the turkey industry, a commonly asked question involves the relationship between age and carcass/breast yield. As toms are grown to similar market weights but at increasingly younger ages, whether breast component yields are similar or whether other age related factors influence the proportional distribution of different carcass parts is an important issue. An experiment was conducted to study this subject. Toms from a common experiment were processed at 16, 17, 18, and 19 weeks of age (approximately 115 to 125 per age) and those toms processed at the latter age (19 weeks) were weighed weekly beginning at 7 days. The body weight data from all four ages were subsequently divided into six classes (< 11.4 kg; 11.4 to 12.3 kg; 12.3 to 13.2 kg; 13.2 to 14.1 kg; 14.1 to 15 kg; > 15 kg). This type of body weight distribution allowed for statistical evaluation of the independent effects of age *versus* body weight on selected carcass characteristics.

The mean body weight from the above experiment increased from 11.7 kg at 16 weeks of age to 14.4 kg at 19 weeks. There were no significant differences in the relative weight of the *P. major* due to processing age (17.5 *versus* 18% respectively) but within the different body weight classes, the relative weight of the *P. major* increased from 17% to 18% (linear effect, P = 0.119) (Figure 5.1).

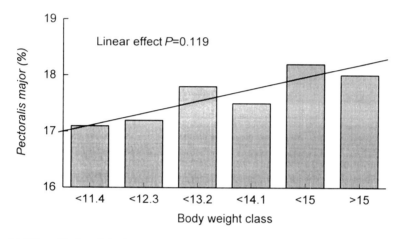

Figure 5.1 Effect of body weight class on relative weight of *Pectoralis major*

This was partially offset, however, by a decline in the relative weight of the *P. minor* with each incremental increase in body weight (4.5 % down to 4.1%) (Figure 5.2).

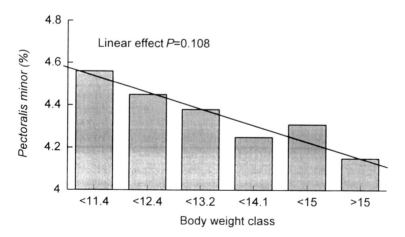

Figure 5.2 Effect of body weight class on relative weight of *Pectoralis minor*

The relative weights of the drum (5.51 %) and thigh (6.75 %) peaked at 16 and 17 weeks of age, respectively, and then declined with age and heavier body weights. Both the drum (tibia) and thigh (femur) are associated with skeletal components as is the *P. minor* (keel), so the relative growth of the muscles associated with the skeleton will decrease with the onset of skeletal maturity. Within each age, the correlations between both live weight and carcass weight and different carcass components were determined. At all four processing ages, the correlations between live or carcass weight and the weight of the *P. major* were significant and similar (r = 0.75 - 0.80) (Figure 5.3).

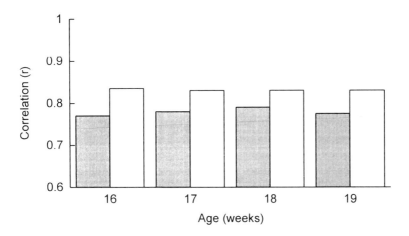

Figure 5.3 Correlation between age and live weight ☐ or carcass weight ☐ for *Pectoralis major*

With the *P. minor*, however, the correlations at 16 weeks (r = 0.30 - 0.40) were considerably lower than at the three older ages (r = 0.65 - 0.70) (Figure 5.4).

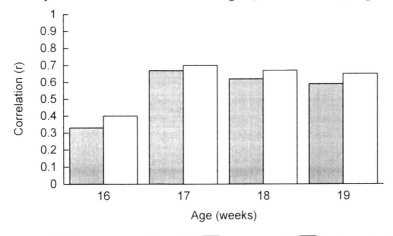

Figure 5.4 Correlation between age and live weight ☐ or carcass weight ☐ for *Pectoralis minor*

This same age trend was also observed for the testes (16 weeks, r = 0.20; 17 - 19 weeks, r = 0.40) (Figure 5.5).

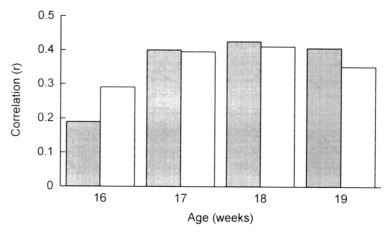

Figure 5.5 Correlation between age and live weight ☐ or carcass weight ☐ for testes

This suggests that in the years to come, as toms are genetically capable of reaching heavier weights at younger ages, factors which negatively or positively influence sexual maturity may need to be incorporated into management practices to achieve the same carcass component yield seen today with older toms of similar body weights.

Those toms which were weighed weekly and processed at 19 weeks were arbitrarily divided into the heaviest 50% (n = 58; top 50) and the lightest 50 % (n = 59; bottom 50).   The mean body weight of the top 50 was 15.27 kg versus 13.59 kg for the bottom 50 (+ 12 %). Tracking these toms back to younger ages showed that the differences were smaller but still significant at 5 weeks (4.2 %) and 10 weeks of age (6.3 %). At processing (19 weeks) age, the differences between groups in the weights of the *P. major* (15.3%), abdominal fat (16.5 %) and testes (49%) were greater than were the differences in body weight suggesting with age, changes in body weight alone will not necessarily influence all carcass components similarly.

With respect to the nutrition of toms, particularly the protein and amino acid needs, there is considerable interest in the amino acid requirements for optimal breast yield independent of that needed for body weight gain. In a 20-week experiment, Lilburn and Emmerson (1993) fed diets containing lysine and total sulfur amino acid concentrations higher than those recommended by the U.S. National Research Council (NRC, 1984) for the first 12 weeks. At the latter age, a random sample of toms from each of two commercial strains fed the higher density diet had increased body weights and *P. major* weights. At 20 weeks, however, when all remaining toms were processed, only one strain continued to show a body weight and breast muscle weight response. Recent work at the University of Arkansas (P. Waldroup, personal communication) has also shown that 105% of the NRC (1994) allowance supported maximal body weight gain, feed efficiency, and breast muscle yield. Data from the literature show that genetic selection for body weight alone has an inconsistent effect at best on the relative weight of the breast (Nestor et al., 1987; Emmerson et al., 1991; Nestor et al., 1995). Within commercial selection programs, however, emphasis on improved conformation in tandem with increased body weight might alter the nutritional requirements for maximal protein accretion.

Most experimentation aimed at defining the amino acid requirements for toms normally includes several replicate pens for each level of the amino acid of interest. This approach for older toms is inherently biased by the normal variation in body weight and feed intake within a group at older ages. In as much as the industry normally uses the practice of feed scheduling, i.e. a defined quantity of feed per bird, it seemed that requirements based on feed intake might worth studying.

At approximately 15 weeks of age, toms from two commercial strains were placed in individual pens with litter floors (24 per strain). A basal diet (3200 kcal/kg ME; 16% C.P.; 0.82% lysine ; 0.70% TSAA) was formulated and ground maize was replaced by supplemental L-lysine to achieve dietary concentrations of 0.82 % (basal), 0.94%, 1.05%, and 1.17%. Feed intake and body weight were calculated and measured weekly and, at 18 weeks, those toms who gained weight weekly were processed. At the start of the experiment, body weight was equalized across treatments and strains and over the course of the study, there were no strain or diet effects on feed intake, body weight gain, eviscerated carcass weight, or *P. major* weight. When feed intake was regressed on body weight gain, the r^2 was significant for both strains (Strain A, 0.675; Strain B, 0.793) (Figure 5.6).

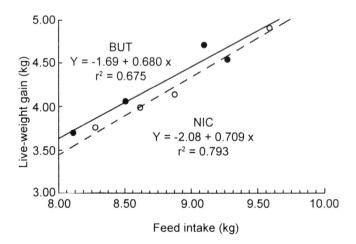

Figure 5.6 Feed intake and live-wight gain 15 to 18 weeks (● BUT, O NIC)

Similar results were observed for the relationship between feed intake and eviscerated carcass weight (Strain A, 0.468; Strain B, 0.561) (Figures 5.7 and 5.8).

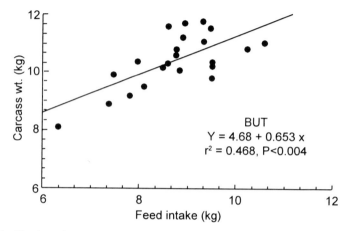

Figure 5.7 Feed intake and carcass weight 15 to 18 weeks (● BUT Toms)

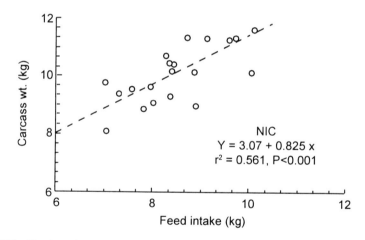

Figure 5.8 Feed intake and carcass weight 15 to 18 weeks (O NIC Toms)

The lysine intake data created in this study suggested that the requirement for maximal gain and eviscerated weight were greater for Strain B toms compared with Strain A toms:

Gain; Strain B, $Y = 1.04 + 35.7X$; $r^2 = 0.529$; $P = 0.0002$;
Strain A, $Y = 3.09 + 12.04X$; $r^2 = 0.195$, $P = 0.039$;   (1)

Carcass Weight; Strain B, $Y = 7.08 + 37.2X$, $r^2 = 0.301$,
$P < 0.01$; Strain A, $Y = 9.26 + 11.67X$, $r^2 = 0.138$, $P = 0.088$).   (2)

The data suggest that although there were no significant mean differences across treatments or strains, the toms within each strain were responding differently to lysine intake. Future studies should address the interactions between energy, protein (amino acid) intake, and carcass development.

References

Emmerson, D.A., Anthony, N.B., Nestor, K.E. and Y.M. Saif. (1991) Genetic association of selection for increased leg muscle and increased shank diameter with body composition and walking ability. *Poultry Science*, **70**:739–745.

Lilburn, M.S. and Nestor, K.E. (1991) Body weight and carcass development in different lines of turkeys. *Poultry Science*, **70**:2223–2231.

Lilburn, M.S. and Emmerson, D.A. (1993) The influence of differences in dietary amino acids during the early growth period on growth and development of Nicholas and British United Turkey toms. *Poultry Science*,**72**:1722–1730.

Nestor, K.E., W.L. Bacon, P.D. Moorhead, Y.M. Saif, G.B. Havenstein and P.A. Renner. (1987) Comparison of bone and muscle growth in turkey lines selected for increased body weight and increased shank width. *Poultry Science*, **66**:1421–1428.

Nestor, K.E., Saif, Y.M., Emmerson, D.A. and N.B. Anthony. (1995) The influence of genetic changes in body weight, egg production, and body conformation on organ growth of turkeys. *Poultry Science*, **74**:601–611.

Waldroup, P.W. (1996) The University of Arkansas. (Personal communication).

IV

Ruminant Nutrition

6

FATS IN DAIRY COW DIETS

P.C. GARNSWORTHY
University of Nottingham, Sutton Bonington Campus, Loughborough, Leics LE12 5RD

Introduction

Dairy cows have evolved an extremely efficient digestive system, principally for utilising the poor quality forages found in their natural diet. The most significant component of this system is the rumen, which is a large fermentation chamber containing billions of microorganisms. These microorganisms break down plant material to provide the energy and protein required for their growth and during this process, they produce volatile fatty acids (VFAs - principally acetate, propionate and butyrate) and ammonia as waste products. The VFAs provide the cow with the major source of energy and carbon skeletons for synthetic processes. The ammonia is converted to urea in the liver and either excreted in the urine or recycled *via* the saliva. Microbial protein is made available to the cow when the microorganisms pass out of the rumen for digestion in the small intestine.

Although moderate levels of performance can be achieved from diets containing only grass or forage, high-producing dairy cows require additional energy and protein from supplementary sources. This is particularly true in early lactation, where appetite is usually reduced to such an extent that the cow cannot eat enough to satisfy nutrient requirements for milk production. Even if body condition is controlled to minimise the reduction in appetite (Garnsworthy and Topps, 1982), diets of high energy concentration are required if cows are to achieve their potential dry matter intake (Jones and Garnsworthy, 1989). Traditional sources of supplementary energy include starchy materials such as cereals, or fibrous materials such as by-products like sugar beet pulp. High starch diets can result in rapid fermentation in the rumen and the low pH induced may inhibit forage-digesting bacteria, which are less tolerant of low pH conditions than amylolytic microorganisms. This results in lowered feed intakes and reductions in the butterfat content of milk. Fibrous supplements do not have such deleterious effects on forage digestion since they are broken down by the same cellulolytic microorganisms. However, the relatively slow rate of digestion of these supplements limits their usefulness because of the physical effects of rumen fill on feed intake.

For diets with very high energy concentrations it is necessary to use fats. These have a gross energy content about twice that of grass and cereals; metabolisable energy is about three times as high and net energy about four times (Table 6.1). When formulating rations, fats have the advantage of creating "space" within the diet, due to their high energy content. This makes it easier, and often cheaper, to satisfy the specifications for other nutrients, such as protein, minerals and vitamins. From a compounder's point of view, fats are additionally attractive since they improve the pelleting qualities of compound feeds and reduce dust. However, dietary fats can have important effects on rumen fermentation and animal responses. These will be reviewed in the current chapter.

Table 6.1 ENERGY CONTENT OF SELECTED FEEDS FOR DAIRY COWS (MJ/KG DM)

	Gross energy	*Metabolisable energy*	*Net energy for milk production*
Grass	18.7	11.2	6.9
Grass silage	20.9	11.0	6.8
Barley grain	18.5	13.3	8.1
Fat Prills	39.0	33.0	26.4
Tallow	39.3	30.5	24.4

Sources: AFRC (1993); MAFF (1990); McDonald, Edwards, Greenhalgh and Morgan (1995); NRC (1989)

Rumen Effects

The fat content of "natural" diets for ruminants is less than 50 g/kg. However, if added fat increases the total fat content of the diet to more than about 100 g/kg, digestive problems can occur.

Rumen microorganisms cannot utilise large quantities of fats. Their normal substrates are carbohydrates and proteins, although limited quantities of fatty acids can be incorporated into microorganisms during cell synthesis. More important than the quantity of fatty acids in the diet is their form, since long-chain unsaturated fatty acids have a detergent effect on bacterial cell walls. Under normal circumstances, the ester linkages of triglycerides are rapidly hydrolysed by bacterial lipases in the rumen. Once released from ester combination, unsaturated fatty acids are subsequently hydrogenated to detoxify them. The result of this is illustrated graphically in Figure 6.1. A typical diet for high-yielding dairy cows might contain twice as much unsaturated fatty acids as saturated. Analysis of rumen fluid shows that this ratio is reversed. The proportion of fatty acids in rumen bacteria that are in the unsaturated form is 0.1 and the proportion in protozoa is 0.35, although the latter may be inaccurate since it is difficult to isolate protozoa from feed particles. Because of these defence mechanisms, the current consumer demand for unsaturated fatty acids in food is difficult to satisfy in ruminant products, compared to the situation in non-ruminants, where inclusion of unsaturated fatty acids in the diet usually increases the proportion of such acids in the carcass.

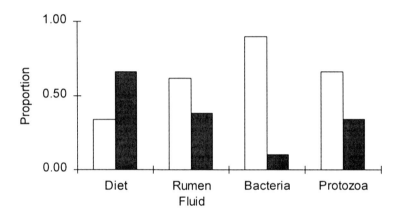

Figure 6.1 Saturation of fats in the rumen (☐ saturated, ■ unsaturated; data from Viviani, 1970)

Fats also have a physical effect on fibre in the rumen; fibre particles become coated with fat, rendering them inaccessible for microbial attack. The magnitude of these effects depends on level, source and type of fat, dietary carbohydrate source and feed intake. A further problem is that long-chain free fatty acids can form soaps with calcium and magnesium in the rumen. This will detoxify the fatty acids, but it can also reduce the availability of the minerals.

Various products have been developed that are referred to as "protected fats". This is really a misnomer since the main requirement is to protect the rumen against detrimental effects of fat, rather than protect the fat against rumen degradation. The term "bypass fat", as used in the US, is mechanistically more accurate in this context. Protected fats offer the potential for increasing the absorption of unsaturated fatty acids from the small intestine. Rumen protection can be natural, (e.g. whole oilseeds, where the slowly digested seed coat slows the rate of fat release), chemical (e.g. encapsulation in formaldehyde-treated casein or formation of calcium soaps) or physical (e.g. selection of fatty acids with a high melting point and small particle size).

Production Effects

The response by dairy cows to supplementary fat is complex and not always predictable. Possible responses that have been reported include increased milk yields, increased or decreased milk fat content, decreased milk protein content, increased live-weight gain and decreased live-weight loss. The observed response will depend on the quantity of fat, its fatty acid profile and degree of protection, the other components of the diet and overall feeding level, the stage of lactation and genetic merit of the cow.

MILK YIELD AND LIVE WEIGHT

Milk yield responses can normally be explained by the increase in total energy intake when fats are given and the increased efficiency of utilisation of energy from fats. If rumen fermentation is disrupted, the response will not be as great as expected, since digestion of the non-fat components of the diet will be reduced. Stage of lactation and genetic merit affect both the milk yield response and the live-weight response. Cows in early lactation, and those of higher genetic merit, partition energy towards milk production at the expense of body fat reserves. Cows normally lose 0.5-1.0 kg of body weight each day for the first eight weeks of lactation and this is mostly body fat. Therefore increased energy intake at this stage of lactation could result in further increases in milk yield if the cow's genetic potential has not been reached and/or a reduction in the daily amount of body fat mobilised. In later lactation, when appetite is greater and milk yield less, partition of energy switches more towards body reserves so increased energy supply results in greater fat deposition in body reserves. Cows that have a high level of body fat reserves respond to high-fat diets by reducing fat mobilisation; cows with a low level of body reserves respond by increasing milk yield (Garnsworthy and Huggett, 1992). Cows of low genetic merit have a greater propensity for fat deposition and will partition a greater proportion of surplus energy in this direction (Holmes, 1988).

MILK FAT CONTENT

Dietary fat can affect milk fat synthesis in a variety of ways and these were summarised by Thomas and Martin (1988), as illustrated in Figure 6.2. If added dietary fats interfere with normal fibre digestion in the rumen, acetate and butyrate production will be reduced and the shortage of precursors in the mammary gland may lead to reduced *de novo* synthesis of milk fat. On the other hand, added fat may increase the quantity of fatty acids available for absorption and secretion in milk. There is also evidence that long-chain fatty acids decrease *de novo* synthesis of fatty acids in the mammary gland even when fed in the protected form so that they do not affect rumen fermentation. In most cases, when dietary fat intake is increased, the proportion of C4-16 fatty acids in milk fat will decrease and the proportion of longer-chain acids will increase. Total milk fat output will depend upon the relative magnitude of the effects on synthesis and absorption, which are affected by protection and degree of unsaturation (Thomas and Chamberlain, 1984).

Unsaturated fatty acids can have double bonds in the *cis* or *trans* form. *Cis* fatty acids have both hydrogen atoms on the same side of the double bond; *trans* fatty acids have one hydrogen atom on each side. Naturally-occurring unsaturated fatty acids contain bonds predominantly in the *cis* form, so it is not surprising that fatty acids in the *trans* form have a much greater inhibitory effect on milk fat concentration than those in the *cis* form. *Trans* fatty acids are present in hydrogenated vegetable oils, but also arise as a result of incomplete biohydrogenation in the rumen. In addition to the general effects of unsaturated fats on fibre digestion, *trans* fatty acids also have a direct inhibitory effect on fatty acid synthesis in the mammary gland (Christie, 1981). Wonsil, Herbein and Watkins (1994) fed diets containing 330 g/kg hydrogenated tallow, menhaden oil or hydrogenated soya bean oil. The menhaden oil did not contain *trans* fatty acids, but was known to affect biohydrogenation in the rumen. The hydrogenated soya bean oil was used as a source of *trans* fatty acids. Hydrogenated tallow did not affect *trans*-oleic acid (18:1) at the duodenum or in milk and did not affect milk fat concentration. Both menhaden and soya oils significantly increased *trans*-18:1 at the duodenum and in milk. They also significantly depressed milk fat concentration (Table 6.2). This suggests that in this experiment the greatest effects of *trans* fatty acids were on milk fat synthesis, rather than fibre digestion in the rumen.

There is currently a great deal of interest in the manipulation of the fatty acid profile of milk fat. This results from consumer demand for unsaturated fatty acids, which are perceived to be more "healthy" than saturated fatty acids. This perception may be based on spurious evidence (Blaxter and Webster, 1992), but the demand is likely to be around for some considerable time yet and warrants attention.

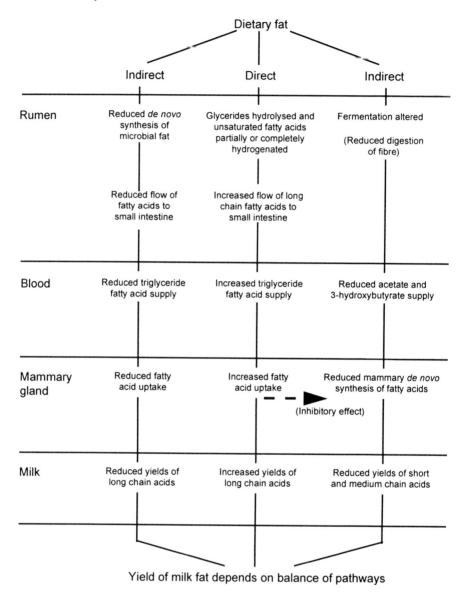

Figure 6.2 Mechanisms by which dietary fat may affect milk fat (after Thomas and Martin, 1988)

It may be thought that the biohydrogenation of fatty acids in the rumen would preclude the presence of unsaturated fatty acids in the milk of cows. However, some fatty acids escape hydrogenation and some are incorporated into microbial lipids. Furthermore, some desaturation of fatty acids takes place in adipose and mammary tissue, so that typically 0.1 of the total fatty acid content of "normal" milk is unsaturated, principally in the form of oleic acid, with small proportions of linoleic. To increase this proportion, it is necessary to use protected fats, either from whole oilseeds or chemical/physical treatment.

Table 6.2 EFFECT OF DIETARY CONCENTRATION AND INTAKE OF *TRANS* FATTY ACIDS ON MILK FAT CONCENTRATION

		Dietary group			
	Control	*Hydrogenated Tallow*	*Menhaden Oil*	*Hydrogenated Soya Bean Oil*	*SEM*
trans-18:1 intake (g/d)	0[a]	12[b]	0[a]	69[c]	2
trans-18:1 flow to the duodenum (g/d)	37[a]	38[a]	163[b]	152[b]	16
milk fat concentration (g/kg)	32.6[a]	31.8[a]	27.8[b]	29.5[b]	0.5
trans-18:1 in milk (g/kg fatty acids)	10[a]	18[a]	134[c]	84[b]	9

[a, b, c] means with different superscripts are significantly different (P<0.05)
Data from Wonsil, *et al.*(1994)

There are numerous reports in the literature describing the effects of different fat sources on milk production, composition and fatty acid profile. Simply increasing the intake of fat can increase the proportion of unsaturated fatty acids in milk fat (e.g. Beaulieu and Palmquist, 1995), because *de novo* synthesis of short-chain fatty acids is reduced and these are all saturated. This makes it difficult to interpret the results of fatty acid profiles from studies designed to investigate a fat supplement (e.g. Garnsworthy and Huggett, 1992) or whole oilseeds (e.g. Bitman, Wood, Miller, Tyrrell, Reynolds and Baxter, 1996), where the comparison is between a low-fat control diet and a high-fat treatment diet.

Where different fat sources have been compared at equal fat intakes, convincing differences in the proportion of unsaturated fatty acids have been found. For example, Schingoethe, Brouk, Lightfield and Baer (1996) compared the effects of extruded soya beans and sunflower seeds on milk fat yield and composition. Neither oilseed increased milk fat content, compared with the low-fat control (Table 6.3), but both increased the proportion of unsaturated acids in milk fat. Interestingly, although soya beans contained nine times more linolenic acid (C18:3) than sunflower seeds, only trace amounts were found in the milk produced from either fat source. It appears that the linolenic acid was hydrogenated to linoleic acid (C18:2) before secretion (Table 6.3).

Table 6.3 MAJOR FATTY ACID COMPOSITION OF MILK FROM COWS FED A LOW-FAT CONTROL DIET AND DIETS CONTAINING EXTRUDED SOYA BEANS OR SUNFLOWER SEEDS (G/KG TOTAL FATTY ACIDS).

	Diet		
	Control	*Soya bean*	*Sunflower*
C16:0[a]	295	215	208
C16:1	41	44	50
C18:0[a]	85	126	136
C18:1[a]	172	278	271
C18:2[b]	23	41	28
Total saturated[a]	716	599	612
Total unsaturated[a]	262	382	369

[a] difference between control and oilseeds $P<0.01$

[b] difference between control and oilseeds $P<0.01$ and difference between oilseeds $P<0.01$

Data from Schingoethe, *et al.* (1996)

Some cereals, because of their high oil content, have a depressing effect on milk fat content that is larger than might be expected simply from the normal effects of starch on rumen acetate:propionate ratios. They also affect milk fatty acid profile. In particular, naked oats (*Avena satura* var nuda) contain almost twice as much lipid as other cereals and a large proportion of this is in the form of C18:1 and C18:2. Fearon, Mayne and Marsden (1996) found that replacing barley with naked oats led to a significant decrease in milk fat content. The milk fat contained significantly higher proportions of long-chain fatty acids, particularly C18:2, which resulted in easier spreadability when the milk was subsequently made into butter. These results can be partially explained by the fact that dietary fat concentrations differed between treatments. The higher fat intake of cows given naked oats would be expected to reduce milk fat content and increase C18 proportions, as discussed above. In addition, the relatively high proportions of C18:1 and C18:2 would be expected to increase the supply of *trans*-fatty acids to the duodenum.

One of the greatest potentials for restoring consumer demand for cows' milk is to emphasise the value of the conjugated linoleic acid it contains. Conjugated linoleic acids (CLAs) are a mixture of *cis*- and *trans*- isomers of linoleic acid with alternating double and single bonds. CLAs have received much attention in the field of human health, since they exhibit anti-carcinogenic properties, reduce atherosclerosis and reduce the ratio of fat to lean in the body. The most abundant natural sources of CLAs are milk and meat from ruminant animals (Chin, Liu, Storkson, Ha and Pariza, 1992). It has been known for some time that CLAs are formed during biohydrogenation of unsaturated fatty acids by rumen microorganisms (Kepler and Tove, 1967), yet there is a dearth of published information on their production. Jiang, Bjoerck, Fonden and Emanuelson (1996) found that a diet with elevated levels of C18:1 increased CLAs in milk fat, particularly when fed at a restricted rate rather than *ad libitum*. However, the effects of C18:1 were confounded with changes in forage to concentrate ratio, so this should only be viewed as a preliminary investigation and further research is required.

Another recent development in the context of human health has been the recommendation that intake of n3 (or omega-3) fatty acids should be increased and that the ratio of n6:n3 fatty acids should be approximately 5:1 (COMA, 1994). The best source of n3 fatty acids is marine oils. These have already been mentioned for their capacity to reduce milk fat concentration due to their propensity to increase the production of *trans*-fatty acids in the rumen (Wonsil, Herbein and Watkins, 1994). Of particular interest are the very long chain fatty acids C20:5 and C22:6, both of which are n3 fatty acids. Ashes, Siebert, Gulati, Cuthbertson and Scott (1992) found that these fatty acids were preferentially deposited in body tissue rather than in milk fat. This presents a problem when trying to increase levels of these fatty acids in milk. In the study of Wonsil, Herbein and Watkins (1994), it was found that although plasma concentrations of C20:5 and C22:6 acids were elevated after supplementation with menhaden oil, these fatty acids were not detectable in milk fat. Mansbridge and Blake (1997), found that C20:5 and C22:6 levels in milk fat were significantly increased when an unspecified fish oil was fed to dairy cows. However, the concentrations were extremely low (4 and 2 g/kg total fatty acids respectively) and the efficiency of incorporation of these fatty acids into milk fat was very poor, with only 0.05 and 0.03 of dietary intake respectively for C20:5 and C22:6 appearing in milk fat.

As stated previously, there are many further comparisons of fat sources available in the literature. Results are quite variable, so it would be useful if the fatty acid profile of milk could be predicted from a knowledge of diet composition. Hermansen (1995) attempted to achieve this by analysing data from 35 published experiments, containing 108 treatments. None of the selected experiments used soaps or fats that were chemically protected against rumen fermentation. The best predictor of short-chain fatty acid content (<C12) was the total fatty acid content of the diet. The best predictors for C12, C14 and C16 acids were their respective concentrations in the diet. For C18:0 and C18:1, the best predictor was the total C18 content of the diet. Prediction of C18:2 and C18:3 content was poor (r^2 = 0.29) when a single predictor (type of fat, i.e. fats and oils *versus* oilseeds) was

used, but better ($r^2 = 0.40$) when the proportion of C18:2 and C18:3 was combined with type of fat. This is not surprising since C18:2 and C18:3 would normally need a high degree of protection to avoid saturation in the rumen and subsequently appear in milk. It is unfortunate that Hermansen (1995) did not extend the analysis to include soaps and other protected fats. Such information would be very valuable when designing diets to produce milk fats tailored for specific niche markets.

MILK PROTEIN CONTENT

The majority of experiments evaluating the use of supplementary fat have found that milk protein content is reduced, although milk protein yield may be unaffected or increased. In some instances, the reduction in milk protein content may be a simple dilution effect (due to increased milk yield), but in most cases it appears to be a genuine physiological response. Wu and Huber (1994) reviewed data from 49 selected trials, involving 83 comparisons between added fat and control diets; it was observed that in most cases milk protein concentration was reduced by the addition of fat (Figure 6.3). The degree of depression seemed to be independent of fat source, although fat prills appeared to be the only fat source with positive effects on milk protein concentration. However, the six data points involving fat prills came from short-term (21 day) crossover experiments, where dry matter intake and milk yield were reduced by fat supplementation.

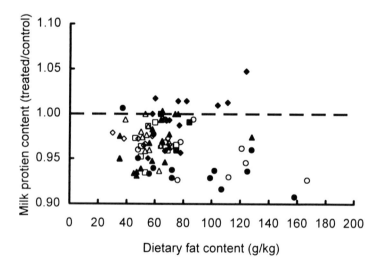

Figure 6.3 Effect of dietary fat concentration on milk protein concentration (■ animal-vegetable blend fat, ▲ calcium salts of fatty acids, ◆ prilled fat, ● protected tallow, ○ tallow, ◇ whole cottonseed, △ whole soyabean, □ yellow grease, - - - control); after Wa and Huber (1994)

The control of milk protein synthesis is a complex process. Dietary protein concentration is only of relevance in deficiency situations and the response is normally determined more by the animal's glucose status. Wu and Huber (1994) discussed four possible mechanisms to explain how dietary fat depresses milk protein concentration; glucose deficiency, insulin resistence, increased energetic efficiency of milk production and somatotrophin deficiency. Of these, glucose deficiency is the strongest contender. Insulin resistence was proposed by Palmquist and Moser (1981), because elevated plasma insulin concentrations have been observed in cows given high-fat diets. However, Wu and Huber (1994) questioned this hypothesis since the effects of supplementary fat on plasma insulin are inconsistent and mammary tissue is not very sensitive to insulin. Increased energetic efficiency was proposed by DePeters and Cant (1992); since the direct incorporation of fatty acids into milk fat with high-fat diets would reduce acetate requirements, availability of glucose would be increased, so mammary blood flow would be decreased, thereby reducing amino acid supply. Wu and Huber (1994) found little data to support this hypothesis. Casper and Schingoethe (1989) suggested that somatotrophin production was inhibited by free fatty acids in the blood, which might reduce mammary uptake of amino acids. However, a subsequent study (Austin, Schingoethe and Casper, 1991) did not support this hypothesis.

Glucose deficiency remains the most likely explanation for the depression in milk protein concentration usually observed with dietary fat supplementation. There are several ways through which fat can affect glucose status.

- Where fat directly replaces starch in a diet, rumen propionate levels may be reduced. Propionate is the major source of glucose in ruminant metabolism.
- Microbial protein synthesis may be lower because of the reduced supply of fermentable metabolisable energy (FME). This would reduce the supply of glucogenic amino acids.
- In ruminants, fats are absorbed from the small intestine as free fatty acids. To convert these to triglycerides for transport, glycerol-3P must be synthesised from glucose present in the cells of the gut mucosae.
- Milk yield often increases after fat supplementation. This implies an increased glucose requirement for lactose synthesis, since lactose is the major osmotic component of milk.

Whether these putative effects of glucose are a direct cause of milk protein depression, or whether they act through a drain on amino acids is not clear. An attempt has been made to calculate the likely magnitude of these changes in glucose requirements, and reduced protein supply *per se*, on the protein status of a dairy cow (Figure 6.4). Some very broad assumptions have been made, but this was necessitated by lack of published data. The conversion of propionate to glucose

Dietary Change

❖ Replace 0.5 kg barley with 0.5 kg calcium-salts of palm acid oil (85% fat)

Assumptions

❖ Metabolisable Energy (ME) and Metabolisable Protein (MP) are currently adequate.
❖ All the response will be in milk production, not body fat reserves.
❖ Conversions of propionate to glucose and glucose to lactose or glyercol-3P have 100% efficiency.

Direct Protein Effects

❖ ME supply increased by 10 MJ/day, sufficient for 2 litres of milk. This increase in milk yield would increase protein requirement by **90g MP**
❖ Fermentable Metabolisable Energy (FME) supply decreased by 6.1 MJ/day. This would decrease microbial protein synthesis by 67 g/day, equivalent to **43g MP**

Glucose Effects

❖ Decreased FME supply reduces propionate production in the rumen by 36 g/day, equivalent to **18 g glucose**
❖ Increased milk yield (2 I/day) contains 90 g lactose, equivalent to 90 g glucose, but it has to be assumed that this is already allowed for by the calculation of increased ME supply to avoid double accounting.
❖ 0.5 kg Ca-salts contains 1.5 M fatty acids. Conversion to triglycerides requires 0.5 M glycerol-3P, equivalent to **47 g glucose**

Indirect Protein Effects

❖ Glucose requirement has increased by 65 g/day, as shown above. If all this was supplied by glucogenic amino acids, it would be equivalent to **117 g MP**

Total increase in protein requirement = 250 g MP

Figure 6.4 Possible changes in metabolisable protein (MP) and glucose requirements when barley is replaced by fat

and glucose to lactose or glycerol-3P are unlikely to be on an equi-molar basis (1.0 efficiency), since some energetic losses would occur during the synthetic processes. However, this assumption causes the magnitude of the effects to be underestimated. The assumption that all of the response to increased energy supply will be in milk production is tenuous. In practice, some shift of partitioning towards body fat reserves is usually observed. Whether this is inevitable, or simply a result of plasma protein/glucose deficiency, is not known. Evidence suggests that increasing the supply of dietary protein to cows with adequate body fat reserves shifts the partitioning of nutrients towards milk production (Garnsworthy and Jones, 1987). Whatever the exact figures are, the conclusion is the same; increasing the fat content of a diet increases glucose and protein requirements. For a cow giving 30 litres of milk per day, the 5% increase in energy supply results in an estimated 15% increase in metabolisable protein requirement. If this requirement is not met, milk protein concentration will decrease or the full potential increase in milk yield will not be realised. It is possible that in some cases the increased milk-protein yield observed with metabolisable protein supplies above theoretical requirements (Newbold, 1994) is due to correction of induced glucose deficiencies caused by high-fat diets.

In order to maintain adequate protein supplies when fats are included in a diet, the dietary protein content should be increased and consideration given to increasing the supply of fermentable carbohydrates. Even where protein supply is more than adequate, glucose status may be marginal with high-fat diets, due to insulin resistance and absorption requirements. This approach was adopted in a series of experiments conducted at Nottingham University. In the first (Garnsworthy, 1996a), the effects of replacing cereals with protected fat and/or lactose was investigated. Lactose was chosen as a source of readily-fermentable carbohydrates since it has less effect on rumen pH than sucrose and might be expected to increase microbial protein yield (Chamberlain and Choung, 1995). Replacing cereals with protected fat led to a significant increase in milk yield and a decrease in milk protein content (Table 6.4). The depression in milk protein content was partially alleviated by inclusion of lactose in the diet. In another experiment (Garnsworthy: unpublished), it was found that an increase in the supply of undegradable protein could maintain milk protein concentrations as milk yields increased in response to protected fat (Table 6.5). In a third experiment, it was found that a combination of strategies produced the best response. When cereals were replaced by a protected fat in combination with protected rapeseed, and lactose was also added to the diet, a significant increase in milk yield was observed without any reduction in milk protein concentration (Table 6.6;Garnsworthy, 1996b).

Table 6.4 MILK YIELD AND COMPOSITION OF COWS GIVEN PROTECTED FAT WITH OR WITHOUT LACTOSE

		Treatment group		
	Control	*Protected Fat*[1]	*Lactose*[2]	*Protected Fat*[1] *& Lactose*[2]
Milk yield (kg/d)	24.6	27.7	25.6	26.5
Milk fat (g/kg)	46.3	46.9	46.8	48.0
(kg/d)	1.13	1.30	1.20	1.27
Milk protein (g/kg)	32.3	30.7	32.7	31.9
(kg/d)	0.79	0.85	0.83	0.85

[1] Calcium salts of palm acid oil
[2] Whey permeate

Source: Garnsworthy (1996a)

Table 6.5 MILK YIELD AND COMPOSITION OF COWS GIVEN PROTECTED FAT WITH PROTEIN IN THE FORM OF RAPESEED MEAL, FISHMEAL OR PROTECTED RAPESEED MEAL

protein source fat source	Rapeseed Megalac[1]	Fishmeal Megalac[1]	MegaPro[2]
Milk yield (kg/d)	26.0	27.1	29.2
Milk fat (g/kg)	49.6	47.6	45.9
(kg/d)	1.29	1.29	1.34
Milk protein (g/kg)	30.0	29.9	29.1
(kg/d)	0.78	0.81	0.85

[1] Calcium salts of palm acid oil
[2] Combined product containing calcium salts of palm acid oil with rapeseed meal added during manufacture (heat-protected)

Source: Garnsworthy, unpublished

Table 6.6 MILK YIELD AND COMPOSITION OF COWS GIVEN COMBINED PROTECTED FAT AND PROTECTED RAPESEED MEAL[1] WITH OR WITHOUT LACTOSE

	Treatment Group			
	Control	Protected Fat + Protein	Lactose	Protected Fat + Protein & Lactose
Milk yield (kg/day)	21.3	23.4	22.7	27.8
Milk fat (g/kg)	42.6	41.7	42.5	41.8
Milk protein (g/kg)	33.7	33.6	33.7	33.8

[1] Calcium salts of palm acid oil combined with rapeseed during manufacture

Source: Garnsworthy (1996b)

Future Possibilities

In the human food industry, dietary trends and nutritional recommendations appear somewhat volatile. It is a feature of affluent societies that consumers want plenty of choice in the foods available to them and most foods are selected for their convenience, rather than nutritional value. However, there are some instances where food fashions have been generated by nutritional considerations and certain people are willing to pay a premium for foods which more closely match their personal "ideal". Whether these nutritional considerations are justified is usually questionable, but there are often niche markets that can be profitably exploited.

One of the longer-lasting, and potentially more damaging, trends is the avoidance or reduction in consumption of animal fats. There is virtually no sound scientific evidence that animal fats harm human health more than vegetable fats. However, a generation has been advised to reduce their consumption of dairy products, with the resulting side-effect of a decrease in calcium intake. The trend towards preference for semi-skimmed or half-fat milk is probably irreversible and farmers are encouraged, through milk-pricing mechanisms, to produce milk with a low fat content. The easiest way to reduce the fat content of milk for liquid sale is to skim off some of the fat during processing. This technology has been used for centuries, but the reduction in demand for butter has reduced the economic value of milk fat. There are simple feeding techniques that can be employed to reduce the butterfat content of milk. Until the early 1980s, dairy farmers were concerned about the so-called "low fat syndrome". A simple method of reducing milk fat concentration is to feed cows on high-starch rations with little forage, which will usually have the additional benefit of maintaining high milk protein concentrations. Unfortunately, this goes against the sound economic advice (in the UK and other forage-growing countries) of producing maximum amounts of milk from home-grown forages. It also tends to increase milk yield, which could be detrimental if a volume quota is operating on milk production.

The use of dietary fat sources that increase the supply of *trans*-fatty acids to the mammary gland, such as naked oats, fish oils and hydrogenated vegetable oils may be a tempting route for farmers who want to reduce the fat content of their milk. However, these fat sources invariably increase the concentrations of *trans*-fatty acids in milk and, as it is now thought that *trans*-fatty acids increase the risk of coronary heart disease (CHD) in humans (Willett, Stampfer, Manson, Colditz, Speizer, Rosner, Sampson and Hennekens, 1993), this may seem to be a retrograde step. It is important to distinguish between *trans*-fatty acids arising from rumen biohydrogenation and those from hydrogenated vegetable oils. *Trans*-C18:1 of ruminant origin has the double bond mainly in the 11-carbon position and does not increase the incidence of CHD; hydrogenated vegetable oils have the double bond mainly in the 9-carbon position, which has been associated with increased risk of CHD (Willett, *et al.*, 1993). This may preclude the use of fats which increase *trans*-fatty acids by direct incorporation, such as hydrogenated vegetable oils, but since the position of the double bond has not been reported in studies with dairy cows, it is not possible to at this stage to draw firm conclusions.

The production of "designer" milk may be one route for adding value to milk destined for niche markets. Of particular interest are n3 fatty acids and conjugated linoleic acid. However, the poor efficiency of incorporation found with n3 acids suggests that this is not a viable option. It would be much more sensible to encourage consumers to eat fish oil directly, rather than utilise an increasingly scarce commodity by feeding it to dairy cows. The study of CLAs has yet to be actively pursued. They do offer potential for enhancing the image of milk but much more information is needed on their occurrence and the factors that affect their concentrations in milk.

References

AFRC (1993) *Energy and Protein Requirements of Ruminants* An advisory manual prepared by the AFRC Technical Committee on Responses to Nutrients. CAB International, Wallingford.

Ashes, J.R., Siebert, B.D., Gulati, S.K., Cuthbertson, A.Z. and Scott, T.W. (1992) Incorporation of n-3 fatty acids of fish oil into tissue and serum lipids of ruminants. *Lipids* **27**: 629–631

Austin,C.L., Schingoethe, D.J. and Casper, D.P. (1991)Influence of bovine somatotropin and nutrition on production and composition of milk from dairy cows. *Journal of Dairy Science* **74**:3920–3932.

Beaulieu, A.D. and Palmquist, D.L. (1995) Differential effects of high fat diets on fatty acid composition in milk of Jersey and Holstein cows. *Journal of Dairy Science* **78**: 1336–1344.

Bitman, J., Wood, D.L., Miller, R.H., Tyrrell, H.F., Reynolds, C.K. and Baxter, H.D. (1996) Comparison of milk and blood lipids in Jersey and Holstein cows fed total mixed rations with or without whole cottonseed. *Journal of Dairy Science* **79**:1596–1602.

Blaxter, K.L. and Webster, A.J.F. (1991) Animal production and food: real problems and paranoia. *Animal Production* **53**: 261–269.

Casper, D.P. and Schingoethe, D.J. (1989) Model to describe and alleviate milk protein depression in early lactation dairy cows fed a high fat diet. *Journal of Dairy Science* **72**:3327–3335.

Chamberlain, D.J. and Choung, J-J. (1995) The importance of rate of ruminal fermentation of energy sources in diets for dairy cows. In *Recent Advances in Animal Nutrition - 1995.* (Eds P.C. Garnsworthy and D.J.A. Cole) pp 3–27. Nottingham University Press: Nottingham.

Chin, S.F., Liu, W., Storkson, J.M., Ha, Y.L. and Pariza, M.W. (1992) Dietary sources of conjugated dienoic isomers of linoleic acid, a newly recognised class of anticarconogens. *Journal of Food Composition and Analysis* **5**: 185–197.

Christie, W.W. (1981) The effects of diet and other factors on the lipid composition of ruminant tissues and milk. In *Lipid Metabolism in Ruminant Animals* (Ed W.W. Christie) pp 193–226. Pergamon Press: Oxford.

COMA [Committee on Medical Aspects of Food Policy] (1994) Nutritional aspects of cardiovascular disease. *Report on Health and Social Subjects* No 46.

DePeters, E.J. and Cant, J.P. (1992) Nutritional factors influencing the nitrogen composition of bovine milk: a review. *Journal of Dairy Science* **75**:2043–2070.

Fearon, A.M., Mayne, C.S. and Marsden, S. (1996) The effect of inclusion of naked oats in the concentrate offered to dairy cows on milk production, milk fat composition and properties. *Journal of the Science of Food and Agriculture* **72**:273–282.

Garnsworthy, P.C. (1996a) The effects on milk yield and composition of incorporating lactose into the diet of dairy cows given protected fat. *Animal Science* **62**:1–3.

Garnsworthy, P.C. (1996b) Response by dairy cows to supplements containing protected fat and protein with and without lactose. *Proceedings of the 47th Annual Meeting of EAAP, Lillehammer, Norway.* **2**:166. Wageningen Pers: Wageningen.

Garnsworthy, P.C. and Huggett, C.D. (1992) The influence of fat concentration of the diet on the response by dairy cows to body condition at calving. *Animal Production* **54**:7–13.

Garnsworthy, P.C. and Jones, G.P. (1987) The influence of body condition at calving and dietary protein supply on voluntary food intake and performance in dairy cows. *Animal Production* **44**:347–353.

Garnsworthy, P.C. and Topps, J.H. (1982) The effect of body condition of dairy cows at calving on their food intake and performance when given complete diets. *Animal Production* **35**:113–119.

Hermansen, J.E. (1995) Prediction of milk fatty acid profile in dairy cows fed dietary fat differeing in fatty acid composition. *Journal of Dairy Science* **78**:872–879.

Holmes, C.W. (1988) Genetic merit and efficiency of milk production by the dairy cow. In *Nutrition and Lactation in the Dairy Cow.* (Ed P.C. Garnsworthy) pp195–215. Butterworths: London.

Jiang, J., Bjoerck, L., Fonden R. and Emanuelson, M. (1996) Occurrence of conjugated *cis* 9, *trans*-11-octadecadienoic acid in bovine milk: effects of feed and dietary regimen. *Journal of Dairy Science* **79**:438–445.

Jones, G.P. and Garnsworthy, P.C. (1989) The effects of dietary energy content on the response by dairy cows to body condition at calving. *Animal Production* **49**:183–191.

Kepler, C.R. and Tove, S.B. (1967) Biohydrogenation of unsaturated fatty acids. III. Purification and properties of a linoleate *cis*-12, *trans*-11 isomerase from *Butyrivibrio fibrisolvens. Journal of Biological Chemistry* **242**:5686–5692.

MAFF (1990) *UK Tables of Nutritive Value and Chemical Composition of Feedingstuffs* Rowett Research Services, Aberdeen

Mansbridge, R.J. and Blake, J.S. (1997) The effect of feeding fish oil on the fatty acid composition of bovine milk. *Proceedings of the British Society of Animal Science,* p22, British Society of Animal Science: Edinburgh.

McDonald, P., Edwards, R.A., Greenhalgh, J.F.D. and Morgan, C.A.(1995) *Animal Nutrition* Longman, Harlow.

Newbold, J.R. (1994) Practical application of the metabolisable protein system. In *Recent Advances in Animal Nutrition - 1994.* (Eds P.C. Garnsworthy and D.J.A. Cole) pp231–264. Nottingham University Press: Nottingham.

NRC (1989) *Nutrient Requirements of Dairy Cattle* National Academy Press, Washington

Palmquist, D.L. and Moser, E.A. (1981) Dietary effects on blood insulin, glucose utilisation and milk protein content of lactating cows. *Journal of Dairy Science* **64:** 1664–1670.

Schingoethe, D.J., Brouk, M.J., Lightfield, K.D. and Baer, R.J. (1996) Lactational response of dairy cows fed unsaturated fat from extruded soybeans or sunflower seeds. *Journal of Dairy Science* **79**:1244–1249.

Thomas, P.C. and Chamberlain, D.G. (1984) Manipulation of milk composition to meet market needs. In *Recent Advances in Animal Nutrition - 1984.* (Eds W. Haresign and D.J.A. Cole) pp 219–245. Butterworths: London.

Thomas, P.C. and Martin, P.A. (1988) The influence of nutrient balance on milk yield and composition. In *Nutrition and Lactation in the Dairy Cow.* (Ed P.C. Garnsworthy) pp97–118. Butterworths: London.

Viviani, R. (1970) Metabolism of long-chain fatty acids in the rumen. *Advances in Lipid Research* **8:** 267–346.

Willett, W.C., Stampfer, M.J., Manson, J.E., Colditz, G.A., Speizer, F.E., Rosner, B.A., Sampson, L.A. and Hennekens, C.H. (1993) Intake of *trans* fatty acids and risk of coronary heart disease among women. *The Lancet* **341:** 581–585.

Wonsil, B.J., Herbein, J.H. and Watkins, B.A. (1994) Dietary and ruminally derived *trans*-18:1 fatty acids alter bovine milk lipids. *Journal of Nutrition* **124**:556–565.

Wu, Z. and Huber, J.T. (1994) Relationship between dietary fat supplementation and milk protein concentration in lactating cows: A review. *Livestock Production Science* **39**:141–155.

7

EFFECTS OF FEEDING STARCH TO DAIRY CATTLE ON NUTRIENT AVAILABILITY AND PRODUCTION

C. K. REYNOLDS, J. D. SUTTON and D. E. BEEVER

Centre for Dairy Research, Department of Agriculture, The University of Reading, Earley Gate, Reading, RG6 6AT, Berks

Introduction

Sustained increases in the genetic potential for milk production by dairy cattle have stressed the need for rations formulated to provide maximal energy and protein availability for milk synthesis. High producing cows produce milk energy at the expense of body energy reserves in early lactation, but the yield achieved and the extent of body tissue mobilization are determined by energy intake relative to the propensity for milk synthesis. By maximizing net nutrient supply from the diet higher yields can be achieved and in the long term the extent of body fat and protein mobilization and associated metabolic and reproductive disturbances may be reduced. High yielding cows have a greater capacity for intake and use of absorbed nutrients and can only achieve their full productive potential through maximal nutrient intake. Maximum energy intake can be achieved by maximizing dry matter (DM) intake and by increasing energy concentration of the diet by adding fat or increasing the inclusion rate of energy concentrates. In practical terms this often means feeding more cereal grain starch. In addition, nutrient availability to the cow can be maximized by simultaneously optimizing rumen fermentation and amounts of protein and carbohydrate digested in the small intestine (Clark *et al.*, 1992). There have been a large number of reviews written on the effect of starch feeding, starch type and processing and starch digestion characteristics on production responses in ruminants (eg. Nocek and Tamminga, 1991; De Visser, 1993; Sutton, 1979 and 1989; Ørskov, 1976 and 1986; Owens *et al.*, 1986; Theurer, 1986) and there is a large volume of literature available on the response of milk fat to starchy concentrate feeding, but there has also been recent interest in the effects of feeding starch on milk protein content. This chapter will review effects of starch supplementation and site of starch digestion on nutrient availability and milk production.

Starch analysis

In spite of the importance of starch as a feedstuff component for all sectors of the animal industry as well as for human diets, it is difficult to obtain precise and accurate estimates of the starch content of ration components. In a recent ring test (Beever *et al.*, 1996) 8 commercial laboratories reported starch content of maize silage samples ranging from 165 to 272 g/kg DM (mean ± sem: 228±36) for a low DM silage (276 g DM/kg) and from 194 to 311 g/kg DM (mean ± sem: 262±42) for a high DM silage (335 g DM/kg). For this test one laboratory refused to report starch concentrations because of their general dissatisfaction with the analytical methods available. Much of the variation between laboratories in the analysis of starch content of feedstuffs relates to differences in the procedures used. There is an official AOAC method approved for measuring the starch content of cereals (AOAC, 1995) which specifies the use of a purified glucoamylase from a specific source under very strict assay conditions, except for the temperature for glucoamylase hydrolysis which is given as 'optimal temperature of glucoamylase used'. Among laboratories using enzymatic assays the procedures published by MacRae and Armstrong (1968) are widely used. Variation between laboratories using assays requiring enzymatic hydrolysis for starch analysis can be attributed to differences in sample preparation, the source of the enzyme used and the conditions used for enzymatic hydrolysis. Sources of enzyme vary considerably in their activity, purity and sensitivity to assay conditions. In addition, batches of enzyme from the same source can vary as well (P. J. Van Soest, personal communication). Sample processing is important as well because gelatinization can destroy carbohydrate and Maillard products can inhibit complete digestion. There are also a number of assays available for total nonstructural carbohydrates (Nocek and Tamminga, 1991), but these measurements often include pectins, fructans, ß-glucans and sugars which have different digestive characteristics than starch. For these reasons, data describing starch digestion from different studies should be compared with caution. Improved methods of starch analysis (De Visser, 1993) or at least a strict standardization of starch analysis procedures are needed for ration formulation based on starch content and digestive characteristics to be justified.

Starch Supplementation, Fibre Digestion and Voluntary Feed Intake

In practice, feeding starchy concentrates often reduces intake of the forage component of the diet such that the forage to concentrate ratio is reduced. However, the response to starch supplementation varies considerably depending on the type of forages and concentrates fed, the method of concentrate inclusion, the physiological status of the cow and resulting effects on intake and production. In cows fed grass silage *ad libitum*, doubling the inclusion of 'standard' starchy concentrates decreased silage dry matter intake by about 0.4 kg/kg concentrate dry matter fed (Sutton *et al.*, 1994), which is similar to responses observed in other studies (Agnew *et al.*, 1996). In contrast, increasing concentrate crude protein content increased silage dry matter intake (Sutton *et al.*, 1994). Therefore any comparison of the digestive and production responses to supplementation of starch with different levels, types or processing characteristics must carefully consider the basal diet employed for the comparison. For example the response to grain fed within a total mixed ration (TMR) is likely to differ from the response achieved when forage is fed *ad libitum* and grain is only provided in large meals fed at milking. When grass silage was fed 2 or 4 times per day or in a TMR and concentrate was fed at 2, 4, 6 or 8 kg/d, concentrate feeding caused a linear decrease in silage intake, but higher silage intakes were achieved with the TMR (Agnew *et al.*, 1996). The substitution rate of silage for concentrate was approximately 0.5, 0.4 and 0.3 for twice daily, 4 times daily and TMR feeding, respectively. Decreases in silage intake with concentrate feeding are probably due to multiple factors such as the total capacity for organic matter digestion and metabolizable energy (ME) utilization as well as specific effects on rumen fermentation and the products of digestion (Forbes, 1995).

As observed by Agnew *et al.* (1996), forage intake is often greater when concentrates are fed in a TMR rather than as meals provided at milking. In addition to the obvious difference in the ability of the cow to discriminate against diet components, and thus alter diet forage to concentrate ratio and total forage intake by selection, changes in the pattern of rumen digestion also contribute to the reduction in forage intake with meal feeding of concentrates. When starchy concentrates are fed as meals the fermentation of large amounts of starch over short periods of time can lead to the production and absorption of large amounts of volatile fatty acids (VFA), with associated increases in ruminal VFA concentration. Even when as much as 90% of the diet is composed of rapidly degraded starchy concentrates, rumen pH concentrations remain fairly constant when the daily ration is provided in frequent (hourly) meals but fall dramatically when concentrates are only fed at milking (Sutton *et al.*, 1986). These differences in the pattern of rumen acid concentration were associated with differences in metabolic and production responses to meal frequency. Feeding at milking alone caused a greater nutrient and insulin elevation in jugular vein blood and a larger depression in milk fat concentration than when large amounts of starch were fed more frequently (Sutton *et al.*, 1985).

Accumulation of acid in the rumen when cereal starches are digested in the rumen can damage the rumen epithelium and inhibit the activity of celluloytic microorganisms (Ørskov, 1976) which can cause a depression in fibre digestion and may lead to reductions in forage and (or) total dry matter intake (Grant, 1994). In general, increasing the rate of starch digestion in the rumen by processing or by feeding more rapidly digested starch types such as barley or wheat causes a greater depression in rumen pH and fibre digestion than when more slowly digested starch types such as maize or sorghum are fed (Ørskov, 1976). In early lactation cows, feeding steam-rolled barley compared to ground maize increased ruminal digestion of starch, decreased dry matter intake and decreased neutral detergent fibre (NDF) digestibility in the rumen and total digestive tract (McCarthy *et al.*, 1989). Although in this study rumen pH, measured hourly, was not different across diets, concentrations of VFA were much higher and pH was numerically lower in rumen fluid when barley was fed. Thus a depression in cellulolytic activity of the rumen microflora may have contributed to the lower digestion of fibre when barley was fed, although rumen pH was very low (less than 6) and both rumen and total tract NDF digestibility were low for all treatments in this study (McCarthy *et al.*, 1989). The low fibre digestibilities measured in these cows may also be attributed to low diet crude protein concentrations, which were less than 150 g/kg DM, and resulting low rumen ammonia concentrations which may also have limited the growth of cellulolytic microbes. Rumen ammonia concentration in these cows was lowered by feeding barley compared with maize, presumably due to a greater use of ammonia for microbial protein synthesis in the rumen, which may also explain the depression in fibre digestion and intake when barley was fed. Similar responses were observed when barley was incrementally substituted for ground maize in TMRs containing over 160 g crude protein/kg dry matter, except that depressive effects of increased barley inclusion on rumen pH and fibre digestion were more significant (Overton *et al.*, 1995). Starch source also may alter fibre digestibility via other factors, such as changes in microbial species and the efficiency of microbial growth, which also influences the efficiency of microbial protein synthesis. Total fibre digestion was also depressed when barley was substituted for maize in mid-lactation cows, but rumen pH and dry matter intake were not affected (DePeters and Taylor, 1985). In contrast, fibre digestion in the rumen and dry matter intake tended to be higher when ground barley was compared to ground maize in studies with lactating dairy cows (Herrera-Saldena *et al.*, 1990). The fibre in these diets was provided primarily by alfalfa hay and cotton seed hulls. In other studies at the same location, increasing rumen digestion of sorghum or maize starch by steam processing increased rumen fibre digestion and (or) dry matter intake (Poore *et al.*, 1993; Chen *et al.*, 1994). These responses may be explained by an enhancement of the digestion of grain fibre itself or a limited availability of rapidly fermentable organic matter for microbial growth when maize or sorghum grain-based diets are fed. It appears then that the intake response to starch supplementation of dairy cow rations is influenced by the method of feeding employed, the level and degradability of dietary protein, the type of forage and the type and amount of starch fed.

In addition to effects on fibre digestion, feeding additional starch may also reduce total dry matter and energy intake via increased ruminal production and absorption of propionate and subsequent metabolic and endocrine responses (Forbes, 1995). In early lactation, forage intake was higher when cows were supplemented with a high fibre concentrate compared to a starchy concentrate providing equal ME (Garnsworthy and Jones, 1993). Rumen fluid from steers fed these concentrates contained higher concentrations of propionate and less acetate when the starchy concentrate was fed. The effect of increased propionate absorption into the portal vein on intake in *ad libitum* fed, lactating dairy cows was clearly demonstrated by the mesenteric vein propionate infusion studies of Casse *et al.* (1994). Cows reduced their dry matter intake during propionate infusions such that the total amount of propionate removed by the liver was remarkably similar to the amount removed during control infusions, but the absorption of other nutrients such as lipogenic precursors was reduced. The reduction of acetate, butyrate and ß-hydroxybutyrate availability without a concomitant drop in glucose supply provides a likely explanation for the decrease in milk fat concentration observed (Casse *et al.*, 1994).

Milk fat responses to starch feeding

Decreased milk fat concentration is the most predictable production response to increased starchy concentrate supply (Sutton, 1989) and the response is generally associated with a shift in VFA production in the rumen away from acetate and towards propionate or perhaps more specifically a reduction in the relative availability of lipogenic versus glucogenic precursors. The reduction in milk fat concentration is generally associated with an increase in lipid deposition in body tissue. This shift in energy partition away from milk towards body tissue is often illustrated by data from calorimetry studies conducted by W. P. Flatt *et al.* in which lactating dairy cows were fed diets differing in the ratio of alfalfa hay to ground maize and soyabean meal (the mixture formulated to be isonitrogenous to the hay) and their energy balance measured (Tyrrell, 1980). As concentrate level increased dry matter intake decreased, total ME remained relatively constant and milk energy output decreased due primarily to reductions in milk fat concentration, but tissue energy retention was increased such that net energy balance (milk plus tissue energy) remained constant. This shift in energy retention from milk towards body fat was associated with a decrease in rumen acetate to propionate ratio and presumably an increase in glucose availability. Other studies have shown that feeding higher concentrate diets increases total glucose turnover and plasma insulin concentrations (Evans *et al.*, 1975). Higher insulin concentrations should depress lipolysis and promote lipogenesis at the expense of milk lipid production, but recent studies have questioned the role of insulin in milk fat depression (McGuire *et al.*, 1995). In lactating dairy cows infusion of large amounts of insulin and glucose for 4 days caused a nonsignificant decrease in milk fat concentration which was considered small relative to the severe milk fat depression often observed when high starch levels are fed. While elevated insulin may not be the sole cause of milk fat depression, a shift in tissue energy balance towards body fat deposition accompanies the depression in milk fat output and this response is associated with, and in part mediated by, an elevation in peripheral insulin concentration.

Recently the role of trans- geometric isomers of unsaturated fatty acids as a potential cause of milk fat depression when high concentrate diets are fed (Davis and Brown, 1970) have received renewed interest (Gaynor *et al.*, 1994; Wonsil *et al.*, 1994; Romo *et al.*, 1994; Griinari *et al*, 1995). Cereal grains contain oils of which a large proportion are often linoleic acid and many maize varieties can be particularly high in oil content (4.1% or more) compared with other cereal grains. Trans-isomers are a step in the microbial biohydrogenation of linoleic acid to stearic acid. If biohydrogenation of linoleic acid is incomplete, or rumen turnover is increased by high intakes, then large amounts of trans-octadecanoates may reach the small intestine. Recent abomasal infusion studies have shown that increased intestinal supply of trans-fatty acids markedly reduces milk fat concentration, probably via direct effects on the mammary gland (Gaynor *et al.*, 1994; Romo *et al.*, 1994). Milk fat depression was caused by the addition of oil containing unsaturated fatty acids to a high (80%) concentrate diet, but not when the oil was added to a 40% concentrate diet (Griinari *et al.*, 1995). The addition of an oil high in saturated fatty acids had little effect on milk fat concentration when added to either basal diet, suggesting that trans-fatty acids cause milk fat depression as a consequence of incomplete biohydrogenation of unsaturated fatty acids when high starch diets are fed. It appears then that when large amounts of starchy concentrates are fed severe milk fat depression may to a large extent be a consequence of increased absorption of specific trans-isomers of unsaturated fatty acids.

Energetic efficiency

The energetic benefit of feeding starch versus cellulose as a dietary carbohydrate for ruminants has long been recognized (Armsby, 1903), but the reasons for the improved efficiency of metabolizable energy (ME) use for production (milk and body tissue synthesis) for concentrates compared to forages has long been the subject of scientific debate. As for milk fat depression, the response is often attributed to a shift in the proportions of acetate and propionate produced in the rumen and subsequent postabsorptive effects on lipogenic and glucogenic nutrient supply and metabolism, with higher efficiency thought to be a consequence of greater availability of glucose relative to acetate (Armstrong and Blaxter, 1961). There is a much larger data base available for comparing the energetic efficiency of concentrates versus forages in growing animals than in lactating dairy cows, but in early calorimetry studies at Beltsville increasing forage to concentrate ratio from 50:50 to 100:0 (by varying proportions of alfalfa hay versus maize and soyabean meal) reduced the efficiency of ME use for milk energy adjusted for tissue energy loss or gain (ie net energy corrected for tissue energy loss) from 0.65 to 0.54 when a constant maintenance requirement was assumed (Coppock *et al.*, 1964). In other work at Beltsville increasing the maize and soyabean meal content of a maize-silage based diet from 300 to 600 g/kg dry matter fed decreased estimated maintenance energy requirement, but the efficiency of ME use for milk energy corrected for tissue energy gain or loss, excess protein intake and pregnancy (ie net energy milk [NE_{milk}]) decreased from 0.61 to 0.54 (Tyrrell and Moe, 1972). This decrease in the use of ME for milk and body tissue energy was in part due to a decrease in milk yield and fat content, but also may have been a consequence of a low diet crude protein content for the high maize meal diet (140 g/kg DM) compared with the low maize meal diet (155 g/kg DM).

The difference in the efficiency of ME utilization between forages and concentrates may to a large extent be attributable to alterations in gut metabolism. In order to achieve equal ME intakes greater dry matter intakes are required for forages, therefore gut fill and the work of rumination and digestion are generally greater than for concentrates. In addition, greater amounts of acetate are produced in the rumen. Greater dry matter intake, gut fill and acetate absorption can all contribute to an increase in gut mass. Ultimately these differences in the mechanical and biochemical processes of carbohydrate digestion and nutrient absorption affect portal-drained visceral (PDV; the gut, spleen, pancreas and associated fat) heat production and the relative outputs of acetate versus glucose from the total splanchnic (PDV and liver combined) tissues (Reynolds *et al.*, 1994). In growing beef heifers fed diets similar to those used in previous Beltsville calorimetry studies (Coppock *et al.*, 1964) differing in forage:concentrate ratio (75:25 versus 25:75) and fed at two equal ME intakes, ME use for tissue energy was lower for the high forage diet due to greater heat production. This difference between diets in the loss of ME as heat was due almost totally to differences in PDV heat production (Reynolds and Beever, 1995). In addition, the PDV of heifers fed the high forage diet absorbed more acetate and used more glucose from arterial blood on a net basis. Liver glucose production was the same for the two diets, but due to differences in absorption the ratio of acetate:glucose released by the splanchnic tissues was reduced by half when the concentrate diet was fed. This difference in acetate availability relative to glucose may also contribute to differences in the efficiency of ME utilization for growth, but it is important to remember that at maintenance intakes as much as half of the acetate produced in the rumen is utilized by the PDV (Bergman and Wolff, 1971).

A number of studies with lactating dairy cows have determined the utilization of energy from diets containing varying types of concentrates. In many of these studies there is little difference between types of concentrates in terms of the efficiency of utilization of ME for NE (milk plus tissue energy), but substantial differences in the extent to which ME is partitioned between milk and tissue. For example, cows fed a fibrous concentrate, beet pulp, retained a greater proportion of ME for milk energy versus tissue energy compared with when they were fed a maize-based concentrate (Tyrrell, 1980). Similar results were obtained when fibrous (sugar beet and citrus pulp with cotton seed) and starchy (barley, wheat and maize gluten) concentrates were compared using grass-silage based diets (Gordon *et al.*, 1995). These differences probably reflect a difference in the relative absorption rate of lipogenic and glucogenic precursors. In addition, studies comparing genotype (Moe and Tyrrell, 1979), maturity (Wilkerson and Glenn, 1996) and processing (Moe *et al.*, 1973; Wilkerson and Glenn, 1996) of maize grain have found that differences in energy utilization are due primarily to effects on organic matter and energy digestibility, but that greater starch digestibility can shift energy partition away from milk and towards body tissue (Moe *et al.*, 1973). However, in studies where physical form or maturity increased grain digestibility and ration metabolizability there were concommitant increases in milk yield and protein content (Moe *et al.*, 1973; Wilkerson and Glenn, 1996).

It is well established that differences in the efficiency of energy utilization or nutrient partitioning between lactation and body tissue arising from the feeding of varying carbohydrate sources are associated with, and probably attributable to, differences in the proportions of glucogenic and lipogenic VFA produced in the rumen. However, data describing the production of any of the VFA in the rumen of lactating dairy cows at high levels of intake are scarce (Sutton, 1985). With one exception (Wiltrout and Satter, 1972) data describing the production of any of the VFA in cows producing more than 20 kg milk/day are nonexistent, yet VFA production rates are the basis of most current models of nutrient utilization in lactating dairy cows. In spite of the difficulties in obtaining these measurements, and the errors of measurement involved, they are sorely needed if for no other reason than to confirm or dispute the assumption that relative concentrations reflect relative production rates (Sutton, 1979). As part of a major effort conducted at the former National Institute for Research in Dairying at Shinfield to determine the effects of starchy concentrate feeding on digestive and production responses in lactating dairy cows (Sutton *et al.*, 1980), a study was conducted in 5 lactating rumen-fistulated Friesian cows fed twice daily diets containing either 60 or 90 % barley-based concentrates with grass hay and supplying equal intakes of digestible energy. Measurements of acetate, propionate and n-butyrate production in the rumen were measured by non-steady state isotope dilution using ^{14}C-acetate, ^{14}C-propionate and ^{14}C-n-butyrate on 3 separate days at the end of each dietary treatment period (Sutton *et al.*, unpublished observations; Table 7.1). The data clearly show a decrease in acetate production and increase in propionate production with increased concentrate intake, but the relative response of propionate production, which more than doubled, is much greater than the relative response of acetate production. These are the only measurements of ruminal production in lactating dairy cows which include all 3 major VFA in the same cows during more than one dietary treatment period. With the current limitations on the conduct of these types of intensive studies it is unlikely that similar measurements of VFA production from high yielding cows will be repeated. Without measurements of rumen production, measurements of net PDV absorption of VFA will provide insight into their availability to tissues other than the PDV, but these measurements underestimate true rates of absorption because PDV tissues utilize VFA.

Table 7.1 NET VFA PRODUCTION (MOL/DAY) IN LACTATING DAIRY COWS FED AT EQUAL DIGESTIBLE ENERGY INTAKE BARLEY-BASED CONCENTRATES AT 2 LEVELS OF INCLUSION WITH GRASS HAY (Sutton *et al.*, unpublished observations; for ration details see Sutton *et al.*, 1980).

| | *Concentrate inclusion* | | | |
	60 % Barley	*90 % Barley*	*SEM*	*P<*
Acetate	56.8	48.3	3.3	0.10
Propionate	16.3	36.4	2.8	0.01
n-Butyrate	6.5	4.7	1.2	NS

Site of starch digestion

Another consideration when feeding starch to lactating dairy cows is the extent to which starch escapes the rumen and is digested in the small intestine or the hindgut.

This has been the subject of numerous reviews since the early work in Kentucky (Karr *et al.*, 1966) and the United Kingdom (Armstrong and Beever, 1969; Ørskov *et al.*, 1969). Chemical and structural characteristics affecting extent and site of starch digestion vary both with source (Waldo, 1973; Ørskov, 1986; Owens *et al.*, 1986) and processing (Hale, 1973; Ørskov, 1976; Theurer, 1986). Of the 4 cereal grains most commonly fed to lactating dairy cows in the United States and United Kingdom, wheat and barley provide more rapidly digested starch to the rumen than maize and sorghum because they have a higher proportion of soluble starch (65 vs 30 %) which is more rapidly digested (21 vs 4 %/h). Based on *in situ* disappearance rates and estimates of rumen outflow in lactating dairy cows, de Visser (1993) calculated that 42 % of insoluble starch in maize and sorghum escapes the rumen, compared to only 8 % for wheat and barley. A commonly held belief is that the energetic efficiency of starch digested in the intestines will be greater than for starch digested in the rumen because small intestinal starch digestion liberates glucose directly while ruminal starch digestion produces VFA, of which the propionate fraction can be converted to glucose in the liver. Although data from glucose infusion studies support this concept in sheep fed at maintenance (Armstrong *et al.*, 1960), production studies generally show a greater efficiency of utilisation for liveweight gain when grain is processed to increase ruminal digestion (Owens *et al.*, 1986). This is because total starch digestion and ME are increased as well. However, multiple regression analysis of data available suggested that starch digestion in the small intestine provides 42 % more energy than starch digested in the rumen of growing cattle. In addition, infusion studies suggest an improved rate of gain when glucose is provided to the abomasum rather than the rumen, but the additional gain was accounted for primarily as increased abdominal fat deposition (Owens *et al.*, 1986). In growing heifers, the partial efficiency of ME use for tissue energy was 0.47 for a barley-based ration and 0.53 for a maize-based ration (Tyrrell *et al.*, 1972).

There is little evidence in the literature that at equal rates of digestion starch digested postruminally is used for milk production more efficiently than starch digested in the rumen. When ground barley and maize were compared in maize silage-based diets fed to lactating dairy cows the proportional use of ME for NE_{milk} was 0.58 for barley-based diets and 0.63 for maize-based diets, but the difference was not significant (Tyrrell and Moe, 1974). A recent review of the effects of site of starch digestion on lactation performance in dairy cows concluded that "production studies provide no clear evidence that site of starch digestion enhances milk yield or changes composition" (Nocek and Tamminga, 1991). Comparisons of starch type and processing methods which alter site of starch digestion in production studies are often confounded by alterations in dry matter intake or changes in total tract starch digestibility (McCarthy *et al.*, 1989; Overton *et al.*, 1995; Khorasani *et al.*, 1994; Poore *et al.*, 1993). In lactating cows fed 65%

concentrates with alfalfa hay and cotton seed hulls, substituting sorghum for barley increased postruminal starch digestion by 1.4 kg/d but decreased total starch digestion by 0.8 kg/d (Herrera-Saldena *et al.*, 1990). In a series of studies conducted at the University of Arizona, milk yield and milk protein composition have been increased by feeding steam flaked sorghum compared with dry rolled sorghum (eg Chen *et al.*, 1994; Poore *et al.*, 1993). In one of these studies, steam flaking sorghum increased starch digestion in the rumen by 1.5 kg/d and decreased postruminal starch digestion by 0.6 kg/d, thus total tract digestion increased by 0.8 kg/d (Poore *et al.*, 1993). In addition, microbial protein synthesis increased at the expense of undigested feed protein flow to the intestines, but the net effect was an increase in total protein availability for digestion. In general, sorghum starch is the least digestible starch source fed to dairy cattle in the United States and shows the most improvement in feeding value for fattening rations when processed to increase its digestibility (Hale, 1973; Ørskov, 1976). Similarly, lactation responses to steam flaking of maize grain are less than for sorghum (Chen *et al.*, 1994).

Numerous studies measuring duodenal starch flow and faecal starch output have shown that the capacity for postruminal starch digestion in the lactating dairy cow is large. In the study of Klusmeyer *et al.* (1991a) postruminal starch digestion reached 5.4 kg/d and averaged over 4 kg/d in a number of recent studies for which ground maize was fed to lactating dairy cows (Table 7.2), but few studies have measured the capacity for small-intestinal starch digestion in the lactating dairy cow. These measurements are needed in order to predict production responses to shifting the site of starch digestion. Total tract digestion of maize is often lower in cattle than sheep, due to differences in grain mastication and flow rate from the rumen, therefore feeding values determined in sheep should be applied to lactating cows with caution (Reynolds and Beever, 1995). In 11 trials with growing cattle fed on maize grain processed by various methods and encompassing 44 treatment means, starch digestibility in the small intestine was 530 g/kg (Owens *et al.*, 1986). In these same trials, total postruminal starch digestibility was 720 g/kg. For the additional starch digestion in the large intestine and caecum the average digestibility was 390 g/kg. Starch digestibility in the rumen for maize-based diets averaged 720 g/kg, thus the digestibility of maize starch in steers appears to decrease as it flows through the gastrointestinal tract. This is because within each successive site of digesion the more digestible starch fraction is removed first, leaving the fraction with lower rate and extent of potential digestion which is more likely to escape to the next site of digestion. With greater intake level and rate of passage for lactating dairy cows compared to steers starch digestibility in the rumen would be expected to be lower and in lactating dairy cows consuming from 3.8 to 10.6 kg ground maize starch/d ruminal starch digestibility averaged 460 g/kg (Table 7.2).

Table 7.2 RECENTLY PUBLISHED MEASUREMENTS OF RUMINAL AND POSTRUMINAL STARCH DIGESTION IN LACTATING DAIRY COWS.

Diet	DM Intake	Milk yield	Starch passage, kg/d			Digestion, g/kg		Reference
			Intake	Duodenal	Faecal	Rumen	Postrumen	
Barley + CSM	18.3	20.0	5.8	1.2	0.3	800	709	Herrera-Saldena et al., 1990.
Barley + BDG	19.6	"	6.0	1.2	0.4	800	681	
Sorghum + CSM	20.2	"	6.3	3.2	0.6	498	800	
Sorghum + BDG	18.3	"	4.5	2.5	0.6	488	744	
Maize SBM - 14.5% CP	21.8	29.3	9.1	3.8	0.5	583	868	Klusmeyer et al., 1990.
Maize SBM - 11.0% CP	20.9	26.9	9.8	4.6	0.7	530	848	
Maize MGM - 14.5% CP	20.9	29.6	9.5	4.4	0.6	530	864	
Maize MGM - 11.0% CP	21.6	26.6	10.4	4.6	0.9	556	804	
Maize SBM	25.1	39.9	10.6	5.4	0.8	485	854	Klusmeyer et al., 1991a.
Maize SBM + CaLCFA	23.8	39.9	9.0	5.4	0.4	399	930	
Maize FM	23.4	41.7	9.8	5.3	0.5	460	879	
Maize FM + CaLCFA	22.3	40.5	8.9	5.2	0.4	417	925	
Low forage	25.5	36.0	9.3	4.7	0.4	499	917	Klusmeyer et al., 1991b.
Low forage + CaLCFA	24.0	37.4	8.0	4.4	0.3	434	922	
High forage	24.7	35.3	7.0	3.5	0.3	494	914	
High forage + CaLCFA	23.3	35.7	5.9	2.8	0.2	525	914	
Ground maize	23.1	31.2	7.6	4.2	0.2	455	943	Cameron et al., 1991.
Ground maize + urea	23.0	31.8	7.5	4.9	0.2	344	946	
Ground maize + starch	21.6	29.9	8.3	4.2	0.2	491	926	
Ground maize + both	21.0	31.4	8.0	3.8	0.1	524	954	

Table 7.2 Continued

Diet	DM Intake	Milk yield	Starch passage, kg/d			Digestion, g/kg		Reference
			Intake	Duodenal	Faecal	Rumen	Postrumen	
Ground maize	19.9	29.1	5.8	3.6	0.3	433	900	Lynch *et al.*, 1991.
Ground maize + Met/Lys	19.4	27.9	5.5	3.3	0.2	432	926	
Ground maize + bST	20.0	33.3	5.8	4.1	0.2	306	933	
Ground maize + both	19.9	31.2	5.8	3.6	0.2	401	939	
Alfalfa + rolled sorghum	20.7	18.2	6.2	3.6	1.0	426	691	Poore *et al.*, 1993.
Alfalfa + flaked sorghum	20.7	19.2	6.1	1.8	0.1	711	921	
Straw + rolled sorghum	21.3	18.6	7.2	3.3	1.4	535	565	
Straw + flaked sorghum	20.7	19.6	6.8	1.6	0.1	763	915	
Maize:Barley - 100:0	22.8	26.9	7.5	4.3	0.7	419	809	Overton *et al.*, 1995.
75:25	22.1	27.8	7.3	2.8	0.5	606	804	
50:50	21.3	26.6	7.1	2.7	0.4	609	800	
25:75	19.5	25.2	6.6	1.7	0.3	744	801	
0:100	19.6	22.6	6.7	1.7	0.3	744	812	
Maize - dry rolled	16.5	22.9	3.9	2.0	0.9	470	488	Plascenia and Zinn, 1996.
Maize - flaked, .39 kg/l	17.8	25.5	4.3	2.3	0.3	453	834	
Maize - flaked, .32 kg/l	17.3	25.6	4.3	1.5	0.1	646	915	
Maize - flaked, .26 kg/l	17.8	25.5	4.8	1.5	0.1	691	958	

Short term infusion studies have suggested that as starch supply to the abomasum increases fractional digestion in the small intestine decreases (Kriekemeier *et al.*, 1991). However, in growing cattle which were adapted to their dietary treatments, regression of rate of starch digestion in the small intestine on rate of starch flow to the duodenum was linear and positive. This suggests that in these studies, where starch flow to the small intestine reached over 4 g/kg BW, the capacity of the small intestine for starch digestion was not exceeded (Owens *et al.*, 1986). For published studies in lactating dairy cattle a positive, linear relationship between starch flow to the intestines and total postruminal starch digestion has also been reported (Nocek and Tamminga, 1991). Their regression suggested that total postruminal starch digestibility was 730 g/kg and that the capacity for digestion was not reduced at duodenal flows of up to 4 kg/d, which agrees with the conclusions of Owens *et al.* (1986) for growing cattle. However, in more recent studies in lactating dairy cows (Table 7.2), total postruminal starch digestion averaged 860 g/kg. For the studies reported in Table 7.2, relationships between starch inputs and ruminal, postruminal and total tract starch digestion are shown in figures 7.1, 7.2 and 7.3, respectively. The linear regression for data from studies where ground maize was the primary starch source is also depicted. The slope of the regression for total postruminal starch digestion for all starch sources (0.945 ± 0.037; $r^2 = 0.95$) suggests that the capacity for postruminal starch digestion in the lactating dairy cow is high. With one exception where ground maize was fed (Plascencia and Zinn, 1996), sorghum appeared to be the only source of starch resistant to postruminal digestion for the studies cited. Nocek and Tamminga (1991) implied that with increasing postruminal supply the digestibility of starch in the small intestine of the lactating dairy cow is decreased and an increasing proportion of postruminal digestion occurs in the hindgut, but rates of starch digestion in the small intestine appear to have been estimated using data from growing cattle (Owens *et al.*, 1986). Starch fermentation in the hindgut provides VFA which are available for absorption, but microbial protein synthesized in the hindgut is not digested and elevates endogenous faecal N (Ørskov, 1986), which decreases apparent N digestibility.

Available data describing small-intestinal starch digestion in lactating dairy cows are presented in Table 7.3. These data were all obtained using markers to estimate starch flow in the duodenum and ileum and in each case problems in obtaining representative digesta samples, or with the markers themselves, were reported. In the study of Knowlton *et al.* (1996) greater flow of starch in the ileum than in the duodenum for rolled maize was blamed on segregation of particles (maize grain) during sampling and processing. Similarly, problems encountered with measurements of duodenal flow using cannulae placed near the pylorus were attributed to segregation of digesta particles at the sampling site and therefore the data reported for starch flow were based on measurements obtained using a more

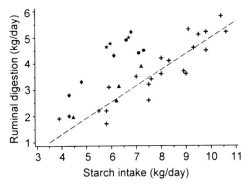

Figure 7.1. Relationship between ruminal starch digestion and starch intake in lactating dairy cows fed diets containing starch primarily from ground maize (✚), barley (★), sorghum (▲), steam-flaked maize or sorghum (◆) or mixtures of ground maize and barley (●). The linear regression for ground maize based diets (y = -1.154 + 0.614(x), r²=0.837) is shown. Data from Table 7.2.

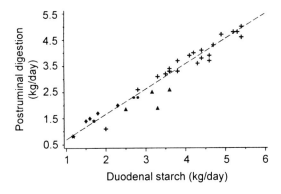

Figure 7.2 Relationship between postruminal starch digestion and duodenal starch flow in lactating dairy cows fed diets containing starch primarily from ground maize (✚), barley (★), sorghum (▲), steam-flaked maize or sorghum (◆) or mixtures of ground maize and barley (●). The linear regression for ground maize based diets (y = -0.272 + 0.965(x), r²=0.918) is shown. Data from Table 7.2.

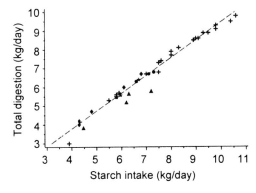

Figure 7.3 Relationship between total starch digestion and starch intake in lactating dairy cows fed diets containing starch primarily from ground maize (✚), barley (★), sorghum (▲), steam-flaked maize or sorghum (◆) or mixtures of ground maize and barley (●). The linear regression for ground maize based diets (y = -0.025 + 0.950(x), r²=0.982) is shown. Data from Table 7.2.

Table 7.3 SITE OF STARCH DIGESTION IN LACTATING DAIRY COWS FITTED WITH RUMINAL, DUODENAL AND ILEAL CANNULAE.

	DMI	Milk yield	Starch passage, kg/d				Starch digestion, g/kg			Reference
			Intake	Duodenal	Ileal	Faecal	Rumen	S. Intestine	Hindgut	
Control	17.9	28.5	4.2	0.5	0.12	0.03	887	745	750	a
+ Fat	16.2	29.0	2.4	0.2	0.06	0.02	901	750	667	
+ Protein	17.2	30.6	2.9	0.2	0.05	0.02	921	780	600	
+ Fat + Protein	16.0	28.8	1.6	0.2	0.03	0.01	893	824	667	
Dry maize - ground	23.4	35.2	7.9	3.1	2.7	0.9	609	132	674	b
Dry maize - rolled	23.4	33.4	7.9	2.5	2.9	2.0	692	-	696	
HM maize - ground	24.4	35.0	8.8	1.1	0.5	0.2	868	588	319	
HM maize - rolled	23.7	35.2	8.3	1.6	0.6	0.4	812	633	345	
Barley - 60%	12.9	14.6	4.0	0.6	0.1	0.1	843	773	562	c
Barley - 90%	11.0	14.5	5.5	0.8	0.2	0.1	857	715	663	
Maize - 60%	12.4	15.6	4.4	2.2	0.6	0.2	507	738	438	
Maize - 90%	12.0	17.5	6.4	3.5	1.1	0.7	454	676	368	

a - Palmquist *et al.*, 1993
b - Knowlton *et al.*, 1996
c - Sutton and Oldham, unpublished

distal cannulation site after the pancreatic duct (Palmquist *et al.*, 1993). The data of Knowlton *et al.* (1996) do suggest that although duodenal starch flow was lower for the less mature, fermented maize grain, due to more extensive digestion in the rumen, the starch reaching the duodenum was more digestible and therefore total starch digestion in the small intestine was greater for the high moisture than the more mature, dry grain. Thus, for the rations based on mature maize, a much greater proportion of total-tract starch digestion occurred in the hindgut.

The data of Sutton and Oldham (Table 7.3) are unpublished data from 4 lactating cows in a study for which the experimental design did not allow direct statistical comparisons, therefore the data presented are treatment means unadjusted for missing observations or period effects. In addition, problems were encountered with the use of the marker employed (Cr_2O_3) and therefore corrections have been applied. These flow rates were measured in cows fed either 60 or 90% concentrates based on rolled barley or ground maize in a 2 × 2 factorial design. Measurements of VFA production for the barley-based diets (Table 7.1) were obtained using different cows as a companion study to these measurements of site of starch digestion. Compared to barley, feeding maize decreased starch digestion in the rumen and increased starch digestion in both the small intestine and hindgut, but the digestibility of starch digested in the small intestine was not dramatically lower for maize (711 g/kg) than for barley (744 g/kg). In contrast, starch digestibility in the hindgut did appear to be lower for maize (408 g/kg) than for barley (612 g/kg), but the amount of starch reaching the hindgut was much lower for the barley-based diets. Across all these studies, the largest amount of starch digested in the small intestine of the lactating dairy cow was 2.4 kg/d (Table 7.3). Total postruminal starch digestibility was 909 and 855 g/kg for barley and maize starches, respectively. These values for maize starch digestion in the small intestine and postruminal tract are higher than the averages reported for growing cattle, but not outside the range (Owens *et al.*, 1986).

In a recent study in early lactation cows, a suspension of maize starch was infused into the duodenum of 4 cows at 0, 700, 1400 and 2100 g/d in a 4 x 4 Latin Square design with 2-week periods for adaptation to the infused starch (Reynolds *et al.*, 1996). Dry matter intake and faecal starch concentration were not affected by starch infusion, thus even at the highest level of infusion nearly all the starch supplied was digested. During the last 4 days of each infusion, milk yield was increased by starch infusion, but milk fat concentration was reduced, especially at the lower rates of infusion (Table 7.4). This reduction in milk fat concentration, although small, was probably caused by an increase in glucose availability and could not be attributed to an increase in trans-fatty acid absorption because trans-fatty acids produced in the hindgut would not be available for absorption. As a consequence of the reduction in milk fat concentration, milk energy output was only increased by the highest level of starch infusion, suggesting that much of ME from the starch infused was either oxidized or used for body tissue energy retention. At the highest level of starch infusion the recovery of infused gross energy as milk energy output was only 17.5%. Although nearly all the starch infused was digested, the proportion of infused starch digested in the small intestine to glucose versus the proportion digested in the hindgut is uncertain. Any carbon from starch fermented in the hindgut could be absorbed as VFA and utilized in the gut or postabsorptive tissues, emitted as gas, or excreted as microbial biomass in faeces.

Table 7.4 MILK COMPOSITION AND YIELD DURING THE LAST 4 DAYS OF 10-DAY DUODENAL INFUSION OF MAIZE STARCH IN 4 DAIRY COWS IN EARLY TO MID-LACTATION (REYNOLDS *et al.*, 1996).

	Starch, g/d					
	0	*681*	*1374*	*2019*	*SEM*	*P<*
DM intake, kg/d	18.0	18.2	18.4	18.4	0.2	0.521
Faecal starch, g/kg	6.8	7.5	11.3	14.5	3.4	0.412
Milk yield, kg/d	31.8	33.1	32.5	33.8	0.3	0.039
FCM, kg/d[1]	32.4	32.7	32.6	33.9	0.1	0.006
Milk energy, MJ/d[2]	97.8	98.8	98.4	102.9		
Protein content, g/kg	29.9	29.8	30.4	30.2	0.4	0.717
Protein yield, g/d	943	986	984	1012	19	0.264
Fat content, g/kg	41.6	39.3	40.1	40.7	0.4	0.071
Fat yield, g/d	1315	1298	1305	1361	4	0.005
Lactose content, g/kg	46.2	46.2	46.2	46.1	0.1	0.959
Lactose yield, g/d	1468	1529	1506	1559	14	0.065

[1] Fat corrected (4%) milk.
[2] Calculated as described by Tyrrell and Reid (1965).

Postruminal starch digestion and glucose absorption

Across a variety of rations, measurements of net PDV glucose flux in lactating cattle vary around zero and in many cases are negative (Reynolds *et al.*, 1994). Even when large amounts of maize grain are fed, net PDV glucose absorption is low relative to the amount of starch estimated to be reaching the small intestine (Weighart *et al.*, 1986; Casse *et al.*, 1994). A negative PDV glucose flux does not mean that no glucose is absorbed, but that PDV use of arterial glucose exceeds glucose absorption into the portal vein. Numerous studies in growing cattle have found that increases in net glucose absorption by the PDV only account for roughly a third of the glucose in starch provided to the small intestine (Harmon, 1992), but this recovery assumes a complete digestion of starch to glucose. In studies where net PDV glucose absorption and changes in ileal flow of starch were measured simultaneously, the recovery of infused maize starch as increased net PDV glucose absorption ranged from 380 (Kreikemeier *et al.*, 1991) to 570 (Kreikemeier and Harmon, 1995) g/kg starch disappearing from the small intestine. The fate of the missing starch carbon is not certain, but a large part of the missing glucose must be utilized for triacylglyceride and fat synthesis or ATP generation in the PDV. The PDV includes mesenteric and omental fat deposits which account for up to 20% of total body fat. In addition, gut epithelial tissue and muscle use glucose as an energy substrate. Measurements of net glucose absorption do not account for any simultaneous increase in uptake of glucose from arterial blood by the PDV when glucose absorption increases the postabsorptive supply of glucose. Thus net PDV glucose absorption underestimates true glucose absorption to the extent that glucose is utilized both during absorption, before reaching venous blood, and from arterial blood. It should be remembered that the PDV receives roughly 40 to 50% of cardiac output in cattle (Huntington *et al.*, 1990), therefore a similar proportion of glucose reaching the heart is immediately returned to the PDV and available for metabolism there. In a recent study in sheep (Cappelli *et al.*, 1993) intraduodenal infusion of glucose increased body glucose turnover by 67% and PDV utilization of arterial glucose by 48%. These increases in body and PDV glucose utilization represented 78% and 20% of the glucose infused, respectively.

As discussed previously, increased PDV energy deposition will contribute to an overall improvement in energetic efficiency, but at the expense of milk energy output. Theoretically, increased glucose absorption does make more glucose available for milk synthesis if the propensity for milk synthesis is greater than the maximal capacity for glucose extraction by the mammary gland (less the gland's glucose requirement for maintenance and growth). One would expect however that except in very early lactation endogenous synthesis is capable of meeting the needs of the mammary gland and that additional glucose from the small intestine will to a large extent be used for tissue energy retention, cause a reduction in endogenous synthesis, or both. Indeed, postruminal or intravenous infusion of glucose in fed lactating dairy cows causes only a slight, if any, increase in milk yield relative to the amount infused (Reynolds *et al.*, 1994). When starch was infused into the duodenum (Table 7.4), increases in lactose yield accounted for only 25 to 81 g/kg glucose equivalents infused.

Although an increase in arterial glucose utilization by the PDV may in part explain the low recovery of abomasally infused starch as increased net PDV glucose absorption, the recovery of abomasally infused glucose as increased net PDV glucose flux is always higher than for infused starch, dextrin or partially hydrolyzed starch (Harmon, 1992). In steers equipped with ileal cannulae cited previously, increases in net PDV glucose absorption accounted for 900 (Kreikemeier *et al.*, 1991) and 733 (Kreikemeier and Harmon, 1995) g/kg of infused glucose not reaching the ileum. This is much higher than the recovery of abomasally infused starch in these steers, therefore either there are fates other than glucose absorption for starch disappearing from the small intestine or the glucose liberated by starch digestion is metabolized differently than infused glucose. This might be due to differences in the postabsorptive endocrine and metabolic response to the glucose absorbed. One major difference between the two routes of carbohydrate administration is the region of the small intestine in which glucose is available for absorption. In multicatheterized steers increases in net PDV glucose flux accounted for a much greater proportion of glucose infused into the duodenum than when glucose was infused into the mid-jejunum and in lambs the mRNA for glucose transporter is higher in the proximal 25% of the intestine (Krehbiel *et al.*, 1996). On the other hand, less than 10 % of starch digestion occurred prior to the mid-jejunum (first 40% of the small intestine) in steers abomasally infused with starch (Russell, 1979 as cited by Krehbiel *et al.*, 1996). Data from growing steers indicates that maltase and isomaltase activity is lowest in the duodenum and highest in the mid-jejunum and ileum in cattle (Kreikemeier *et al.*, 1990). Thus ruminants appear to be efficient at absorbing glucose from the proximal small intestine, but inefficient at digesting starch to glucose prior to the mid-jejunal region where glucose absorption is severely limited. This increases the concentration of glucose and oligosaccharides in the ileum where increasing pH and back flush from the large intestine may allow a limited microbial population to survive and ferment small amounts of carbohydrate (Armstrong and Beever, 1969). In addition, the presence of undigested carbohydrate and VFA may stimulate release of ileal peptides (Taylor, 1993) which may also modulate metabolic responses to starch via signals which are distinct from those arising when abomasally infused glucose is absorbed prior to the ileum.

Limits to starch digestion in the small intestine

Although regression analysis has suggested that the digestibility of maize starch in the small intestine remains relatively constant across a broad range of duodenal inputs (Owens *et al.*, 1986), numerous studies in cattle have suggested that the capacity of the small intestine for starch digestion may be limited (Ørskov, 1986; Harmon, 1992). In some cases these limitations may be a consequence of limited treatment periods which do not allow time for adaptation to increased starch supply (Waldo, 1973). Limits to starch digestion may be a consequence of starch physical structure or the capacity of α-amylase, disaccharidases or glucose absorption (Kreikemeier and Harmon, 1995). Both physical processing and maturity of maize grain may alter the accessibility of maize grain starch to enzymatic hydrolysis in the small intestine (Hale, 1973; Knowlton *et al.*, 1996). Data of Krehbiel *et al.* (1996) suggest that glucose-transporter activity limits glucose absorption in the distal small intestine. A resulting increase in luminal glucose concentration may cause end-product inhibition of the activity of disaccharidases directly and α-amylase directly or indirectly via a build up of oligosaccharides (Kreikemeier and Harmon, 1995). The presence of large amounts of ethanol-soluble α-glucoside (amylose) and short-chain α-glucoside (oligosaccharide) in the ileal digesta of steers abomasally infused with starch suggests that the activity of both groups of enzymes may be limiting. Although a number of early infusion studies showed that increasing starch flow to the duodenum increased α-amylase secretion from the pancreas, these studies are confounded by effects of total ME supply (Harmon, 1992). Steers fed maize-based rations had less α-amylase capacity in the pancreas and intestine than steers fed alfalfa hay at equal ME, but α-amylase capacity increased with increasing ME for both rations (Kreikemeier *et al.*, 1990). This suggests that ME, rather than rumen escape starch *per se*, increases α-amylase capacity in the intestine of cattle; however, the positive effect of duodenal protein flow on pancreatic α-amylase secretion must also be considered in interpreting these studies. The presence of similar amounts of short-chain α-glucoside in the ileal digesta of steers abomasally infused with maize starch or maize dextrin suggests that the α-1,4 glucosidase (maltase) may be more limiting than α-1,6 glucosidase (isomaltase) because dextrin is highly branched (Kreikemeier and Harmon, 1995). If the α-1,6 glucosidase was limiting then one would expect more short-chain glucosidase to be present in the ileal digesta of steers infused with dextrin. The limited time for α-amylase action in the proximal small intestine, the limited availability of disaccharidases in the proximal small intestine, the limited capacity of glucose transporters in the distal small intestine and the negative effect of duodenal starch flow on pancreatic α-amylase secretion all contradict the concept that rumen-escape starch is a more efficient source of ME in cattle than rumen degradable starch. Thus, when considering site of starch digestion, we must also consider site of starch digestion within the small intestine.

Milk protein concentration

Recent articles in the popular press have reported that providing more rumen-escape starch by feeding maize will increase milk protein concentration, but on the same page reports from other studies suggest that feeding more rumen-fermentable carbohydrate will increase milk protein concentration (Farmer's

Weekly, 1996). How can these conflicting results be resolved? The small, positive response of milk protein concentration to ME supply is well documented (Emery, 1978) and it is now dogma that increasing ME will increase milk protein concentration if the increased energy is provided via carbohydrate and not fat. This relationship has been found in other summarizations as well (Sutton, 1989), but is not necessarily predictable within individual studies. The response may be due to changes in rumen fermentation as well as changes in postabsorptive metabolic and endocrine responses, particularly changes in glucose and insulin status. With increasing ME from carbohydrate, rates of organic matter digestion in the rumen (fermentable ME [FME]) will in most cases increase as well. As long as sufficient quantities of rumen degradable protein and nonprotein N are available, this will increase microbial protein synthesis in the rumen and metabolizable protein supply to the small intestine (Waldo, 1973). This is due to effects of increased FME and not starch fermentation *per se*. Comparisons between barley and beet pulp have shown that both sources of FME are equally effective in increasing milk protein concentration (de Visser, 1993). Thus a part of the milk protein response to ME must be attributable to increases in FME and microbial protein synthesis, but this cannot explain a milk protein response to increased postruminal starch digestion.

In addition to increasing metabolizable protein supply, increasing ME and FME may also increase glucose supply via increased availability of propionate and other glucose precursors. Based on the rumen infusion studies of Rook and Balch (1961) the milk protein response to ME has often been attributed to an increase in propionate absorption, but few subsequent infusion studies have supported this view. This is in part due to the negative effects of propionate on intake which often confound interpretation of responses (Casse *et al.*, 1994). If increased ME does increase propionate absorption and/or glucose turnover then in most cases this may elicit an insulin response, especially in cows in positive energy balance, and there is evidence that milk protein concentration and yield may be increased with increased glucose and insulin turnover. In early lactation, cows fed rations containing either 20% or 60% maize-based concentrate at equal digestible energy, plasma glucose concentration and turnover during the postprandial period were nearly doubled by feeding the higher concentrate ration and this was associated with a dramatic increase in milk protein concentration (Evans *et al.*, 1975). This increase in glucose availability was associated with an increase in plasma concentrations of insulin. Recent data provide direct evidence that elevated insulin and glucose can increase milk protein concentration (McGuire *et al.*, 1994). Cows were infused intravenously with large amounts of insulin and glucose for 4 days using a euglycaemic insulin clamp technique in which the rate of glucose infusion was varied in order to maintain basal blood glucose concentration during the insulin infusion. Initial trials showed that this would increase milk protein concentration

(McGuire *et al.*, 1995), therefore the study was repeated with the addition of an abomasal casein infusion to increase amino acid supply to the mammary gland. When additional metabolizable protein was combined with the insulin and glucose infusion there was a significant increase in milk yield, milk protein concentration and milk protein yield (McGuire *et al.*, 1994). The mechanisms behind this response are not certain and may involve other hormones in the somatrophic axis. Depending on the physiological status of the cow and her basal diet, ME supply and/or increased ruminal or postruminal starch digestion could all contribute to an increase in glucose and insulin turnover. Cows in early lactation may be resistant to effects of insulin and may use additional glucose for milk synthesis, therefore positive effects of increased glucose and insulin supply may be greater in cows in positive energy balance and with lower somatotropin:insulin ratios. However, any milk protein response will still require an adequate supply of amino acids to the mammary gland.

Based on these observations, elevated glucose supply represents one mechanism by which increased postruminal starch digestion might increase milk protein concentration. It has also been hypothesized that increased glucose absorption might spare amino acids from catabolism in the gut by supplying additional energy for gut metabolism (de Visser, 1993). To date this hypothesis has not been rigorously tested, but the principal amino acids providing energy for gut enterocytes are the nonessential amino acids glutamine, glutamate and aspartate. These are the amino acids present in the largest amount in microbial protein and any sparing of amino acid catabolism by absorbed glucose in the small intestine would probably provide more of these amino acids rather than essential amino acids limiting milk protein synthesis (Reynolds *et al.*, 1995). In cows at peak lactation fed a diet containing 180 g/kg crude protein, duodenal starch infusion had no affect on milk protein concentration or yield (Reynolds *et al.*, 1996), which does not support the concept that increased glucose supply from postruminal starch digestion increases milk protein concentration by changing gut metabolism of amino acids or insulin status.

Conclusions

In general, when lactating dairy cows are fed increasing amounts of cereal starch, ME supply will increase if dry matter intake is not reduced. This is often associated with a decrease in milk fat concentration and an increase in body tissue fat synthesis which may be due in part to a glucose and insulin response, but in more severe cases there is strong evidence that an increased supply of trans-fatty acids is the cause. At the same time it is generally accepted that increased ME from carbohydrate often increases milk protein concentration. This response may be due to an increase in microbial protein synthesis attributable to increased FME in the rumen, but there is evidence that the response may also be linked to an increase in glucose supply and associated changes in insulin status or other hormonal responses. Thus stage of lactation and genetic merit may affect the response. In terms of feeding supplemental starch *per se*, the milk protein concentration response varies considerably. If rapidly digested starch is fed the response may be negative due to depressions in dry matter and total ME intake and resulting decreases in microbial protein and glucose supply. Feeding rumen escape starch reduces rumen acid load and allows more total starch intake, but the recovery of starch reaching the intestines as net glucose absorption is low, in part due to limits to starch digestion in the small intestine and the use of glucose by the PDV, thus a portion of the starch reaching the intestines will be fermented in the hindgut. Fermentation in the hindgut will not benefit microbial protein supply. Therefore, although sources of rumen-escape starch also provide undegraded feed protein to the small intestine the overall effect is usually a decrease in metabolizable protein supply. Any increase in milk protein output will require a sufficient amino acid supply to support that response, and microbial protein may have advantages over many cereal grain proteins in terms of amino acid profile.

References

Agnew, K.W., Mayne, C.S. and Doherty, J.G. (1996) An examination of the effect of method and level of concentrate feeding on milk production in dairy cows offered a grass silage-based diet. *Animal Science*, **63**, 21–32.

AOAC (1995) AOAC Official method 979.10: Starch in Cereals. In *Official Methods of Analysis of AOAC International*, Chapter 32, p 25. Edited by P. Cunniff. AOAC International: Arlington, VA.

Armsby, H.P. (1903) *The Principles of Animal Nutrition*. John Wiley and Sons: New York, NY.

Armstrong, D.A., Blaxter, K.L. and N. McC. Graham (1960) Fat synthesis from glucose by sheep. *Proceedings of the Nutrition Society*, **19**, xxxi–xxxii.

Armstrong, D.A. and Blaxter, K.L. (1961) The utilization of the energy of carbohydrate by ruminants. In *Symposium on Energy Metabolism*, pp 187–199. Edited by E. Brouwer and A.J.H. van Es, European Association for Animal Production Publication Number 10, Wageningen, The Netherlands.

Armstrong, D.A. and Beever, D.E. (1969) Post-abomasal digestion of carbohydrate in the adult ruminant. *Proceedings of the Nutrition Society*, **28**, 121–131.

Beever, D.E., Cammell, S.B. and Humphries, D.J. (1996) The effect of stage of maturity at   the time of harvest on the subsequent nutritive value of maize silage fed to lactating dairy cows. CEDAR Report No 66.

Bergman, E.N., and Wolff, J.E. (1971) Metabolism of volatile fatty acids by liver and portal-drained viscera in sheep. *American Journal of Physiology*, **221**, 586–592.

Cameron, M.R., Klusmeyer, T.H., Lynch, G.L., Clark, J.H. and Nelson, D.R. (1991) Effects of urea and starch on rumen fermentation, nutrient passage to the duodenum, and performance of cows. *Journal of Dairy Science*, **74**, 1321–1336.

Cappelli, F.P., Seal, C.J. and Parker, D.S. (1993) Portal glucose absorption and utilization in sheep receiving exogenous glucose intravascularly or intraduodenally. *Journal of Animal Science*, **70 (Suppl. 1)**, 279.

Casse, E.A., Rulquin, H. and Huntington, G.B. (1994) Effect of mesenteric vein infusion of propionate on splanchnic metabolism in primiparous Holstein cows. *Journal of Dairy Science*, **77**, 3296–3303.

Chen, K.H., Huber, J.T., Theurer, C.B., Swingle, R.S., Simas, J., Chan, S.C., Wu, Z. and Sullivan, J.L. (1994) Effect of steam flaking of corn and sorghum grains on performance of lactating cows. *Journal of Dairy Science*, **77**, 1038–1043.

Clark, J.H., Klusmeyer, T.H. and Cameron, M.R. (1992) Microbial protein synthesis and flows of nitrogen fractions to the duodenum of dairy cows. *Journal of Dairy Science*, **75**, 2304–2323.

Coppock, C.E., Flatt, W.P. and Moore, L.A. (1964) Effect of hay to grain ratio on utilization of metabolizable energy for milk production by dairy cows. *Journal of Dairy Science*, **12**, 1330–1338.

Davis, C.L. and Brown, R.E. (1970) Low-fat milk syndrome. In *Physiology of Digestion and Metabolism in the Ruminant*, pp. 545–565. Edited by A.T. Phillipson. Oriel Press: England.

DePeters, E.J. and Taylor, S.J. (1985) Effects of feeding corn or barley on composition of milk and diet digestibility. *Journal of Dairy Science*, **68**, 2027–2032.

De Visser, H. (1993) Characterization of carbohydrates in concentrates for dairy cows. In *Recent Advances in Animal Nutrition - 1993*, pp 19–38. Edited by P.C. Garnsworthy and D.J.A. Cole. Nottingham University Press: Nottingham.

Evans, E., Buchanan-Smith, J.G., MacLeod, G.K. and Stone, J.B. (1975) Glucose metabolism in cows fed low- and high-roughage diets. *Journal of Dairy Science*, **58**, 672–677.

Emery, R.S. (1978) Feeding for increased milk protein. *Journal of Dairy Science*, **61**, 825–828.

Farmers Weekly. 1996. Milk special. Higher quality - by design. March 22 issue, p. 50.

Forbes, J.M. (1995) Voluntary intake: a limiting factor to production in high-yielding dairy cows? In *Breeding and Feeding the High Genetic Merit Dairy Cow*, pp 13–19. Edited by T.J.L. Lawrence, F.J. Gordon and A. Carson. British Society of Animal Production Occasional Publication No. 19.

Garnsworthy, P.C. and Jones, B.P. (1993) The effects of dietary fibre and starch concentrations on the response by dairy cows to body condition at calving. *Animal Science*, **57**, 15–22.

Gaynor, P. J., Erdman, R.A., Teter, B.B., Sampugna, J., Capuco, A.V., Waldo D.R.and Hamosh, M.(1994) Milk fat yield and composition during abomasal infusion of *cis* or *trans* octadecenoates in Holstein cows. *Journal of Dairy Science*, **77**, 157–165.

Gordon, F.J., Porter, M.G., Mayne, C.S., Unsworth, E.F. and Kilpatrick, D.J. (1995) Effect of forage digestibility and type of concentrate on nutrient utilization by lactating dairy cattle. *Journal of Dairy Research*, **62**, 15–27.

Griinari, J.M., Bauman, D.E. and Jones, L.R. (1995) Low milk fat in New York Holstein herds. In *Proceedings of the Cornell Nutrition Conference for Feed Manufacturers*. pp. 96–105.

Grant, J.J. (1994) Influence of corn and sorghum starch on the in vitro kinetics of foragefiber digestion. *Journal of Dairy Science*, **77**, 1563–1569.

Hale, W.H. (1973) Influence of processing on the utilization of grains (starch) by ruminants. *Journal of Animal Science*, **37**, 1075–1080.

Harmon, D.L. (1992) Dietary influences on carbohydrases and small intestinal starch hydrolysis capacity in ruminants. *Journal of Nutrition*, **122**, 203–210.

Herrera-Saldena, R., Gomez-Alarcon, R., Rorabi, M. and Huber, J.T. (1990) Influence of synchronizing protein and starch degradation in the rumen on nutrient utilization and microbial protein synthesis. *Journal of Dairy Science*, **73**, 142–148.

Huntington, G. B., J. H. Eisemann, and J. M. Whitt. 1990. Portal blood flow in beef steers: comparison of techniques and relation to hepatic blood flow, cardiac output and oxygen uptake. *Journal of Animal Science*, **68**, 1666–1673.

Karr, M.R., Little, C.O. and Mitchell, G.E. (1966) Starch disappearcnce from different segments of the digestive tract of steers. *Journal of Animal Science*, **25**, 652–654.

Khorasani, G.R., De Boer, G., Robinson, B. and Kennelly, J.J. (1994) Influence of dietary protein and starch on production and metabolic responses of dairy cows. *Journal of Dairy Science*, **77**, 813–824.

Klusmeyer, T.H., McCarthy, R.D.Jr., Clark, J.H. and Nelson, D.R. (1990) Effects of source and amount of protein on ruminal fermentation and passage of nutrients to the small intestine of lactating cows. *Journal of Dairy Science*, **73**, 3526–3537.

Klusmeyer, T.H., Lynch, G.L., Clark, J.H. and Nelson D.R. (1991a) Effects of calcium saltsof fatty acids and protein source on ruminal fermentation and nutrient flow to duodenum of cows. *Journal of Dairy Science*, **74**, 2206–2219.

Klusmeyer, T.H., Lynch, G.L., Clark, J.H. and Nelson D.R. (1991b) Effects of calcium salts of fatty acids and proportion of forage in diet on ruminal fermentation and nutrient flow to duodenum of cows. *Journal of Dairy Science*, **74**, 2220–2232.

Knowlton, K.F., Glenn, B.P. and Erdman, R.A. (1996) Effect of corn grain maturity and processing on performance, rumen fermentation, and site of starch diegestion in early lactation diary cattle. *Journal of Dairy Science*, **79 (Suppl. 1)**, 138.

Krehbiel, C.R., Britton, R.A., Harmon, D.L., Peters, J.P., Stock R.A. and Grotjan, H.E. (1996) Effects of varying levels of duodenal or midjejunal glucose and 2-deoxyglucose infusion on small intestinal disappearance and net portal glucose flux in steers. *Journal of Animal Science*, **74**, 693–700.

Kreikemeier, K.K., Harmon, D.L., Peters, J.P., Gross, K.L., Armendariz, C.K. and Krehbiel, C.R. (1990) Influence of dietary forage and feed intake on carbohydrase activities and small intestinal morphology of calves. *Journal of Animal Science*, **68**, 2916–2929.

Kreikemeier, K.K., Harmon, D.L., Brandt, R.T., Avery, T.B. and Johnson, D.E. (1991) Small intestinal starch digestion in steers: effect of various levels of abomasal glucose, corn starch and corn dextrin infusion on small intestinal disappearance and net glucose absorption. *Journal of Animal Science*, **69**, 328–338.

Kreikemeier, K.K. and Harmon, D.L. (1995) Abomasal glucose, maize starch and maize dextrin infusions in cattle: Small intestinal disappearance, net portal glucose flux and ileal oligosaccharide flow. *British Journal of Nutrition*, **73**, 763–772.

Lynch, G.L., Klusmeyer, T.H., Cameron, M.R., Clark, J.H. and Nelson, D.R. (1991) Effects of somatotropin and duodenal infusion of amino acids on nutrient passage to duodenum and performance of dairy cows. *Journal of Dairy Science*, **74**, 3117–3127.

MacRae, J.C. and Armstrong, D.G. (1968) Enzyme method for determination of α-linked glucose polymers in biological materials. *Journal of the Science of Food and Agriculture*, **19**, 578–581.

McCarthy, R.D., Klusmeyer, T.H., Vicini, J.L., Clark, J.H. and Nelson, D.R. (1989) Effect of source of protein and carbohydrate on ruminal fermentation and passage of nutrients to the small intestine of lactating cows. *Journal of Dairy Science*, **72**, 2002–2016.

McGuire, M.A., Griinari, J.M., Dwyer, D.A. and Bauman, D.E. (1994) Potential to increase milk protein in well-fed cows. In *Proceedings of the Cornell Nutrition Conference for Feed Manufacturers*, pp 124–133.

McGuire, M.A., Griinari, J.M., Dwyer, D.A. and Bauman, D.E. (1995) Role of insulin in the regulation of mammary synthesis of fat and protein. *Journal of Dairy Science*, **78**, 816–824.

Moe, P.W., Tyrrell, H.F. and Hooven, N.W. (1973) Physical form and energy value of corn grain. *Journal of Dairy Science*, **56**, 1298–1304.

Moe, P.W. and Tyrrell, H.F. (1979) Effect of endosperm type on incremental energy value of corn grain for dairy cows. *Journal of Dairy Science*, **62**, 447–454.

Nocek, J.E. and Tamminga, S. (1991) Site of digestion of starch in the gastrointestinal tract of dairy cows and its effect on milk yield and composition. *Journal of Dairy Science*, **74**, 3598–3629.

Ørskov, E.R., Fraser, C. and Kay, R.N.B. (1969) Dietary factors influencing the digestion of starch in the rumen and small and large intestine of early weaned lambs. *British Journal of Nutrition*, **23**, 217–226.

Ørskov, E.R. (1976) The effect of processing on diegestion and utilization of cereals by ruminants. *Proceedings of the Nutrition Society*, **35**, 245–252.

Ørskov, E.R. (1986) Starch digestion and utilization in ruminants. *Journal of Animal Science*, **63**, 1624–1633.

Overton, T.R., Cameron, M.R., Elliot, J.P., Clark, J.H. and Nelson, D.R. (1995) Ruminal fermentation and passage of nutrients to the duodenum of lactating cows fed mixtures of corn and barley. *Journal of Dairy Science*, **78**, 1981–1998.

Owens, F.N., Zinn, R.A. and Kim, Y.K. (1986) Limits to starch digestion in the ruminant small intestine. *Journal of Animal Science*, **63**, 1634–1648.

Palmquist, D.L., Wesibjerg, M.R. and Hvelplund, T. (1993) Ruminal, intestinal, and total digestibilities of nutrients in cows fed diets high in fat and undegradable protein. *Journal of Dairy Science*, **76**, 1353–1364.

Plascencia, A. and Zinn, R.A. (1996) Influence of flake density on the feeding value of steam-processed corn in diets for lactating cows. *Journal of Animal Science*, **74**, 310–316.

Poore, M.H., Moore, J.A., Eck, T.P., Swingle, R.S. and Theurer, C.B. (1993) Effect of fiber source and ruminal starch degradability on site and extent of digestion in dairy cows. *Journal of Dairy Science*, **76**, 2244–2253.

Reynolds, C.K., D.L. Harmon and M.J. Cecava (1994) Absorption and delivery of nutrients for milk protein synthesis by portal-drained viscera. *Journal of Dairy Science*, **77**, 2787–2808.

Reynolds C.K. and Beever, D.E. (1995) Energy requirements and responses: a UK perspective. In *Breeding and Feeding the High Genetic Merit Dairy Cow*, Edited by T.J.L. Lawrence, F.J. Gordon and A. Carson, pp 31–41. British Society of Animal Production Occasional Publication No. 19.

Reynolds, C., Crompton, L., Firth, K., Beever, D., Sutton, J., Lomax, M., Wray-Cahen, D., Metcalf, J., Chettle, E., Bequette, B., Backwell, C., Lobley, G. and MacRae, J. 1995. Splanchnic and milk protein responses to mesenteric vein infusion of 3 mixtures of amino acids in lactating dairy cows. *Journal of Animal Science*, **73 (Suppl. 1)**, 274.

Reynolds, C.K., Beever, D. E., Sutton, J.D. and Newbold, J.R. (1996) Effects of incremental duodenal starch infusion on milk composition and yield in dairy cows. *Journal of Dairy Science*, **79 (Suppl. 1)**, 138.

Romo, G., Erdman, R., Teter, B. and Casper, D.P. (1994) Potential role of trans fatty acids in diet induced milk fat depression in dairy cows. *Proceedings of the Maryland Nutrition Conference for Feed Manufacturers*. pp 64–71.

Rook, J.A.F., and Balch, C.C. 1961. The effects of intraruminal infusions of acetic, propionic and butyric acids on the yield and composition of the milk of the cow. *British Journal of Nutrition*, **15**, 361–369.

Sutton, J.D. (1979) Carbohydrate fermentation in the rumen - variations on a theme.*Proceedings of the Nutrition Society*, **38**, 275–282.

Sutton, J.D., Oldham, J.D. and Hart, I.C. (1980) Products of digestion, hormones and energy utilization in milking cows given concentrates containing varying proportions of barley or maize. In *Energy Metabolism, Proceedings of the 8th Symposium*, pp 303–306 Edited by L.E. Mount. European Association of Animal Production Publication 26, Butterworths: London.

Sutton, J. D. (1985) Digestion and absorption of energy substrates in the lactating cow. *Journal of Dairy Science*, **68**, 3376–3393.

Sutton, J.D., Broster, W.H., Napper, D.J. and Siviter, J.W. (1985) Feeding frequency for lactating cows: Effects on digestion, milk production and energy utilisation. *British Journal of Nutrition*, **53**, 117–130.

Sutton, J.D., Hart, I.C., Broster, W.H., Elliott, R.J. and Schuller, E. (1986) Feeding frequency for lactating cows: effects on rumen fermentation and blood metabolites and hormones. *British Journal of Nutrition*, **56**, 181–192.

Sutton, J.D. (1989) Altering milk composition by feeding. *Journal of Dairy Science*, **72**, 2801–2814.

Sutton, J.D., Aston, K., Beever, D.E. and Fisher, W.J. (1994) Milk production from grass silage diets: the relative importance of the amounts of energy and crude protein in the concentrates. *Animal Science*, **59**, 327–334.

Taylor, I.L. (1993) Role of peptide YY in the endocrine control of digestion. *Journal of Dairy Science*, **76**, 2094–2101.

Theurer, C. B. (1986) Grain processing effects on starch utilization by ruminants. *Journal of Animal Science*, **63**, 1649–1662.

Tyrrell, H.F. and Reid, J.T. (1965) Prediction of the energy value of cow's milk. *Journal of Dairy Science*, **48**, 1–9.

Tyrrell, H.F., Moe, P.W. and Oltjen, R.R. (1972) Energetics of fattening heifers on a corn vs. barley ration. *Journal of Animal Science*, **35 (Suppl. 1)**, 277.

Tyrrell, H.F. and Moe, P.W. (1972) Net energy value for lactation of a high and low concentrate ration containing corn silage. *Journal of Dairy Science*, **55**, 1106–1112.

Tyrrell, H.F. and Moe, P.W. (1974) Net energy value of a corn and a barley ration for lactation. *Journal of Dairy Science*, **57**, 451–458.

Tyrrell, H.F. (1980) Limits to milk production efficiency by the dairy cow. *Journal of Animal Science*, **51**, 1441–1447.

Waldo, D.R. (1973) Extent and partition of cereal grain starch digestion in ruminants. *Journal of Animal Science*, **37**, 1062–1074.

Wieghart, M., R. Slepetis, J. M. Elliot, and D. F. Smith. 1986. Glucose absorption and hepatic gluconeogenesis in dairy cows fed diets varying in forage content. *Journal of Nutrition*, **116**, 839–850.

Wilkerson, V.W. and Glenn, B.P. (1996) Energy balance in early lactation Holstein cows fed corn grain harvested as dry or high moisture and ground or rolled. *Journal of Animal Science*, **74 (Suppl. 1)**, 270.

Wiltrout, D.W., and Satter, L.D. (1972) Contribution of propionate to glucose synthesis in the lactating and nonlactating cow. *Journal of Dairy Science*, **55**, 307–317.

Wonsil, B.J., Herbein, J.H. and Watkins, B.A. (1994) Dietary and ruminally derived *trans*-18:1 fatty acids alter bovine milk lipids. *Journal of Nutrition*, **124**, 556–565.

8

FEEDING AND MANAGING HIGH-YIELDING DAIRY COWS — THE AMERICAN EXPERIENCE

CARL E. COPPOCK

Coppock Nutritional Services, Laredo, Texas 78045, USA

Introduction

It was but a few years ago that average annual milk yields of 10,000 kg/cow were achieved only by the best managed herds; today yields of 13,500 kg and more are becoming widespread. New individual milk production records seem to be set nearly every year. Just recently, a new world milk record was set in Wisconsin with 28,804 kg of milk produced in 365 days by a cow milked twice daily. Walton (1983) predicted that by the year 2000, top individual milk production records would approach 32,000 kg per cow per year, and the best herd averages would approach 16,000 kg per cow per year. Although in 1983 this prediction seemed unduly optimistic, it now seems like a near certainty. Although the Holstein breed dominates the dairy cow population in the U.S. (94% of the cows), the Jersey breed is making strong progress. Recently it was noted (Halladay, 1996) that one Jersey herd with the 5th highest production in the U.S., averaged 8,550 kg milk, 420 kg fat and 325 kg of protein per year, considerably higher than the U.S. average for Holsteins (7,530 kg).

For at least a decade, the genetic progress in Holsteins for milk has been about 118 kg per year or 1180 kg for the 10-year interval. In Jerseys, the genetic change for milk was estimated to be 76 kg annually from 1960 to 1987, but 137 kg from 1983 to 1987 (Nizamani and Berger, 1996). An important advantage of the pervasive genetic merit for high milk yields is that dairy cows are highly responsive to good environment in the broad sense of that word. In more markets, price incentives for low somatic cells encourage better control of udder infections; more precise engineering of milking systems and environmental management systems have reduced these constraints. Development of repartitioning agents especially bovine somatotropin (BST) now focus attention on the constraints associated with nutrition and feeding management which often now limit higher but economic yields of milk. The objectives for this paper are to describe several key features of nutrition and management in high-yielding U.S. dairy herds and suggest several areas for improved efficiency.

Dairy management practices

The U.S. is fortunate to now have a National Animal Health Monitoring System which includes a major emphasis on dairy cows. In May, (Anonymous, 1996), Part I: Reference of 1996 Dairy Management Practices, was published and this document contains a large array of data concerning the management and conditions of U.S. dairying. Twenty states participated, representing 83.1% of the U.S. milking cows. The data were obtained from 2,542 dairy producers who were surveyed with an on-farm questionnaire to provide a representative sample. A few of the results are shown in Table 8.1 and others will be referred to in subsequent sections.

Table 8.1 EXAMPLES OF U.S. DAIRY INFORMATION AND MANAGEMENT PRACTICES[a]

Record-Keeping System	Percent Operations	Percent Dairy Cows
Hand written (as ledger)	80.7	73.3
Dairy Herd Improve. (DHIA)	43.4	54.6
Computer at the location	15.1	36.9
Computer — off location	9.9	13.2
Other system	6.0	5.1
Any system	100.0	100.0
Identification Type		
Ear Tags (all types)	81.2	87.3
Collars	22.3	16.3
Photograph or sketch	17.4	10.3
Branding (all methods)	4.9	12.3
Implanted electronic ID	0.3	0.2
Tattoo (other than brucellosis)	6.5	7.8
Other	10.1	6.4
None	8.8	2.5

[a]Part I: Reference of 1996 Dairy Management Practices (NAHMS) Anonymous, (1996)

Feed sources

Although corn (maize), sorghum, barley and oats continue to be dominant feed grains for cattle and soyabean meal and cottonseed meal are important plant protein supplements, many alternative feeds, sometimes called byproducts, and more recently and correctly called co-products are used in cattle feeds in the U.S. A regional bulletin (Bath *et al.*, 1980) listed 357 tabular entries with 49 feedstuffs discussed individually. Many are well known, including molasses, brewers' grains, hominy feed, wheat bran and corn distillers' grains. Others are less well known including almond hulls, apple pomace, buckwheat middlings, caproco, peanut skins, cow peas, vetch seeds and many more. Most are available only in specific regions, for example, California, and some only during certain seasons. Probably no more than 10 to 15 are consistently available in most of the U.S. Two national alternative feeds symposia have been held (1991 and 1995) to address the issues of these feedstuffs in the rations of dairy and beef cattle.

Grasser *et al.* (1995) surveyed an array of industry representatives to evaluate the importance of 9 major by-products used in livestock feeding in California in 1992. These are listed in Table 8.2 with their individual tonnage and market value at that time. California is now the No. 1 milk producing state in the U.S. and these 9 by-products accounted for more than 2.5 million tonnes and about 27% of all concentrates used in California that year. Whole cottonseed was the most important by-product studied and it accounted for 31% of the total tonnage and it provided about 66% of the total crude protein (CP) and 53% of the total net energy for lactation (NE_l) of the 9 by-products. Moreover, cottonseed constitutes about 62% of the total cotton crop yield (lint plus seed). It has become such a respected and sought after feed that now nearly 78% of the California cottonseed crop is fed to cattle, with 20% crushed for oil and the remainder is used for planting seed. Although it is grown across the southern U.S., it is valued so highly that it is shipped into all of the major dairy states on the northern tier. Brewers grains are used exclusively (100%) for cattle feed. It was determined that these 9 by-products could have provided the CP or NE_l for more than 31% of the milk produced in California during 1992.

A nationwide survey of feedstuffs fed to lactating dairy cattle was recently completed by Mowrey and Spain (1996). Dairy nutritionists from 28 states responded and the results were grouped into 5 regions, midwest (MW), northeast (NE), northwest (NW), southeast (SE), and southwest (SW). The predominant concentrate energy feed was corn followed by barley, sorghum, oats and wheat. The primary protein supplements were soyabean meal, cottonseed meal, soyabean seed, and canola (rapeseed) meal. The predominant by-products were whole cottonseed, soyabean hulls, wheat milling by-products and field corn milling by-products. Alfalfa (hay plus silage) was the principal forage fed, followed by corn

Table 8.2 NINE IMPORTANT BY-PRODUCTS USED IN CALIFORNIA IN 1992[a]

By-product	Total as fed tons (1000)	Percentage of total Concentrate (%)	Market Price ($/ton)	Total Value $ (1000)
Almond hulls	498	5.19	66	32,881
Beet pulp	267	2.78	132	35,191
Brewers grains (wet)	409	4.26	35	14,318
Citrus pulp, (17% DM)	336	3.50	13	4,372
Citrus pulp, (30% DM)	91	0.95	35	3,182
Corn gluten feed, (wet)	15	0.15	57	829
Corn gluten meal, (dry)	16	0.17	358	5,853
Whole cottonseed	814	8.48	154	125,356
Rice bran	127	1.33	77	9,802
Total	2,573	26.81		231,786

[a] Grasser *et al.*, (1995)

forage (silage) and grass forage (hay plus silage) and 15 other forages. Probably the majority of U.S. dairy producers believe that to achieve top production, some alfalfa hay or silage must be part of the forage programme. A recent article (Merrill, 1996) described a forage program without either alfalfa or corn silage in which production averaged 10,900 kg per year. A combination of grasses (primarily stored) provided the forage for this herd.

As part of a symposium which addressed management for herds to produce 13,620 kg of milk per cow per year, Jordan and Fourdraine (1993) sent survey forms to 128 producers who had been identified as high milk producing herds, and from 61 surveys returned, found that corn silage was the dominant forage fed followed by legume hay, legume silage, and grass hay. The average producer supplemented these forages with 6.7 different feed additives (sodium bicarbonate, yeast, tallow, niacin, minerals, etc) and 3.5 different alternative feeds (cottonseed, distillers grains, blood meal, fishmeal, etc). As part of the same symposium, Chase (1993) reviewed some of the practices and applications of using the NRC (1989) recommendations for high-yielding cows. In general, the nutrient guidelines were in relatively close agreement with the rations fed in herds with high production.

Several examples of rations fed to high-producing herds in Wisconsin are shown in Table 8.3 (Gunderson, 1992). A wide array of feedstuffs is apparent, in concert with the high dry matter intake (DMI) necessary to support such intense lactation.

Table 8.3 SOME RATIONS OF HIGH-YIELDING HERDS IN WISCONSIN WITH PRODUCTION >11,350 KG/COW PER YEAR[a]

Ingredient	A	B	Dairy C	D	E	F
Hay						1.6
Haylage	9.9	10.0	8.5	11.3	5.5	7.9
Corn silage		1.9	3.5		5.5	
High-moisture shelled corn		6.5	7.2		6.0	9.4
High-moisture ear corn	7.5			7.0		
High-moisture barley					2.9	
Distillers' grains		0.4	0.1			0.5
Wet brewers grains			2.6			
Corn gluten meal	0.5	0.3	0.3	0.3	0.2	
Liquid molasses		0.1				
Soyabeans (roasted)	3.5	0.3	2.2	0.9	2.4	0.5
Soyabeans (raw)				1.3		
Soyabean meal	0.4	1.3	1.0		1.0	
Soyabean meal (expeller)		0.6				0.5
Fish meal	0.4				0.5	
Whole cottonseed		2.1		2.1		3.4
Meat & bone meal	0.5			1.0	0.3	
Blood meal	0.2	0.1		0.2		0.4
Tallow	0.5	0.5		0.3		
Urea (46%)	0.09		0.06		0.16	
Sodium bicarbonate			0.16		0.23	
Dicalcium phosphate	0.25	0.10	0.39	0.12	0.25	0.25
Limestone		0.11	0.09	0.02	0.16	0.17
White salt	0.11	0.04	0.11	0.08	0.11	0.11
Magnesium oxide	0.05	0.03		0.04	0.04	0.05
Zinc methionine	0.005		0.005		0.005	
Micromins/Vits	0.04	0.02	0.04	0.02	0.04	0.06
Yeast			0.11		0.11	

[a]Gunderson, 1992; High group rations, kg dry matter per day.

Feed management systems

Forty years ago pasture was a dominant forage for most dairy herds in the U.S., followed by hay and silage. Nearly all dairies fed concentrates in the barn at the time of milking so that eating occurred during milking. In the late 1950's and early 1960's it was discovered that many lactating dairy cows would respond to additional concentrates above the feeding standards used then. As herd sizes increased during the 1960's there was an increased emphasis on labour efficiency and a corresponding increase in the construction of milking parlours. But as production increased there was also greater mechanization in the parlour which further reduced the time cows spent there to be milked. Even when additional time was provided and precision equipment was operated carefully, there was no way to ensure that the appetite of the cow would induce her to consume the required concentrate. Most parlour concentrate feeding then became free choice feeding for two 10-minute intervals per day. And the resulting "slug feeding" of highly soluble protein and carbohydrate must have been highly disruptive to a ruminal fermentation system which functions best as a steady state system.

The resolution of the parlour concentrate feeding dilemma has taken several forms: 1) feed some concentrate outside the parlour separate from or blended with forage; 2) use computer controlled concentrate feeders (CCCF); 3) feed all of the concentrate outside the parlour either through the computer feeders or blended with forage as a totally mixed ration (TMR). The TMR system of feeding has been reviewed by Coppock *et al.* (1981); and Spain (1995). A TMR is identified as a quantitative blend of all dietary ingredients, mixed thoroughly enough to prevent separation and sorting, formulated to specific nutrient concentrations and offered *ad libitum*. The national survey (Anonymous, 1996) showed that although among all operations only 35.6% fed a TMR, but among the operations with 200 or more cows, 83.5% fed a TMR. Muller (1992) notes that both research and farmer experiences have shown that from 450 to 900 kg or more increases in milk production per cow per year, when herds were switched to a well formulated TMR.

The CCCF was developed to resolve the dilemma of parlour concentrate feeding and under some conditions it has been quite successful. Its greatest application is in smaller herds and where pasture is a dominant forage programme. The feeders may restrict the cow's rate of eating and the proportion of the 24-hr allotment which a cow can receive at any one meal. Some allow 25% of the day's allowance to be eaten during each 6-hr quadrant of the day. But the CCCF deals only with the concentrate portion of the ration; forage(s) must be dealt with by another system. The problem of feeding concentrates separately from forage is that some cows seem to prefer concentrates to forage and other cows vice versa, and it is difficult to ensure that a cow will eat her needed forage if she is given a large allotment of concentrates which is increasingly necessary as higher production becomes the norm. Moreover, a few cows never learn to use the feeders and social dominance may cause disturbances around the feeders. The feeders are relatively expensive and require routine maintenance and frequent resetting as cows' requirements change.

There are at least three cogent reasons for feeding different TMR rations within the milking herd: 1) The cows can be fed rations which are formulated so that their *ad libitum* consumption will result in more closely meeting the cows' nutrient requirements than some average formulation; 2) Feed costs can usually be reduced by feeding rations of lower nutrient concentration to the lower producing cows; 3) There will be less transfer of certain nutrients (e. g., nitrogen, phosphorus and sodium) to the manure by using formulations tailored to requirements, primarily milk production.

Ideally cows would calve in close calendar proximity so that they could remain in the same group for their entire lactation and the diet composition could be adjusted gradually as lactation advanced. This is most feasible for first lactation cows whose pregnancies were synchronized when they were heifers and who reside in large herds. But most cows will need to change groups as they advance in lactation in order to prevent overcondition and to reduce feed costs. Albright (1983) recommends that farmers should "Move small groups of cows. . . Not only is there social pressure on the cow in her new group, but she may have different amounts of feed, a new milker and a different milking time. Try to keep group size stable and no larger that 100 cows".

Frequency of milking

The Dairy Records Processing Laboratory in North Carolina (Butcher, 1996) reported that 757 herds out of a total of 11,557 herds (6.5%) milked their cows 3 times per day (3X). However, the 3X herds averaged 334 cows compared to an all-herds average of 113 cows. Therefore, 19.4% of the cows whose records are processed by that centre are milked 3X. It is not known whether this number is representative of the whole U.S., but because many larger herds are in the southwest and west, it is probably a low estimate.

DePeters *et al.* (1985) used 38 multiparous cows and 15 primiparous cows (Holsteins) in full lactation studies to compare the effects of twice-daily milking (2X) with three times daily milking (3X) on production, and reproductive performance. The older cows milked 3X produced 15% more milk during the complete lactation and the young cows produced 6% more milk during their first lactation. In both groups, neither dry matter nor energy intakes were affected by milking frequency, but neither group gained as much weight during the lactation as their herdmates milked 2X. Reproductive performance of the cows was not affected. These authors emphasized that herds milked 3X successfully will need careful nutrition and reproductive management.

From a study of 28 California herds (average size, 537), Gisi *et al.* (1986) compared the response of these herds to 2X milking for 3 to 17 months, and after a switch to 3X milking for 36 months. In California during 1984, 11% of the herds and 15% of the cows were being milked 3X. In this study, milk production of all herds increased by 12% above that when they were milked 2X. First lactation cows increased their production by 14%, but the range in response among herds was -2 to +32%. It was further noticed that most of the increased response occurred during the first 3 months that the herds were milked 3X. It was emphasized that if increased response to 3X milking is to be sustained over a long term, better nutrition, especially improved forage and perhaps increased feeding frequency, may be essential. It is not known how soon after a 3X milking regimen appetite will increase. The increase in feed energy which is necessary to support a 15% increase in milk production is shown in Table 8.4. For 27 kg of milk, about 10% more energy is required of the same ration. There is no reason to assume an increased metabolic efficiency, so the increased milk production must be supported with greater dietary nutrient intake.

Table 8.4 EFFECT OF A 15% INCREASE IN MILK YIELD ON FEED REQUIRED TO MAINTAIN BODY ENERGY STATUS

	Initial yield		(a 15% increase)
		(kg/day)	
	30.0		34.5
		NE_l (MJ/day)	
Net energy required for[a] :			
Maintenance[b]	40.58		40.58
Growth (+10%)	4.06		4.06
Milk	86.61		99.58
Total NE_l required	131.25		144.22
Difference		12.97	
Percentage increase in feed required[c]		9.88	

[a] From NRC (1989).
[b] Calculated for 600 kg second lactation, nonpregnant cow, giving milk of 35 g/kg fat.
[c] Assumes no change in digestive or metabolic efficiency and a diet of uniform composition.

A study from Michigan (Speicher *et al.*, 1994) shows that the effects of BST and 3X milking are largely additive, although primiparous cows responded more to the combination of BST and 3X milking than did the multiparous cows. If both of these factors are imposed simultaneously, dairy producers should be prepared for the large increase in appetite which will eventually occur.

Bovine Somatotrophin (BST)

BST has been used commercially in the U.S. for nearly 3 years. The national survey (Anonymous, 1996) showed that of all operations only 9.4% used BST, but in herds of 100-199 cows, 18.5% used this product, and in herds with >200 cows, 31.9% had used BST. So resistance to use was greatest in the small herds of <100 cows. And the percentage of cows that are receiving BST is considerably larger than 10%. Use of BST causes an increase in milk yield within 2-4 days of its first use, although maximal response usually takes from 4 to 6 weeks of sustained use. One survey showed that dairymen were getting an average increase of 4.8 kg of milk per cow per day from BST use. At today's relatively high milk prices, this additional milk is worth about $1.68. With a product cost of about $0.40 per cow per day, and increased feed required of about 2 kg ($0.35) and extra labor of $0.10, the return is highly favorable. Studies with the respiration chambers at USDA-Beltsville, MD (Tyrrell *et al.*, 1988), showed that neither digestive nor metabolic efficiency was changed when BST was used. Therefore, the increased milk which occurs with BST use must be paid for with increased feed. Additional milk of 4.8 kg will require about 2 kg more of feed DM of a well balanced TMR. There is no free lunch with the use of this product. One feature of a cow's appetite, is that there is a lag of 4 to 6 weeks after the first BST injection, before the appetite increases for the additional feed. During this lag period, the nutrients to pay for the additional milk must come from body stores, feed, or a combination of the two. Feed nutrients must be available when the appetite increases so that cows have the chance to replenish nutrient reserves as lactation advances. As with 3X milking, some dairy producers have felt that the sharp response to the initial use of BST has not been sustained, but it is likely in these cases, that the dietary nutrients were not available to sustain the increased milk yields which occurred.

So cows respond to BST in 3 ways; a) they produce more milk; b) after a lag period they eat more feed; and c) when they eat more feed, they produce more heat. Therefore, use of BST affects 2 primary management systems: a) nutritional management, and b) environmental management.

Cornell workers (Van Amburgh *et al.*, 1996) have suggested that because of the increased persistency of cows treated with BST, it may be economically desirable and feasible to deliberately extend the calving interval to as much as 18 months. These workers used nine herds to address this subject with some cows assigned to treatments which allowed extended calving intervals and some cows were never rebred, but all treatment cows received BST beginning at 63 days postpartum. In a preliminary report of this study, it was found that as lactation advanced, milk yield response to BST increased, so that an important difference in persistency occurred. Profitability was greater by nearly $0.75 per cow per day for cows that had an 18-month calving interval vs. those with a 13.2-month interval. This increase in profitability occurred because of greater persistency, fewer postpartum metabolic problems and less culling with fewer replacements required. If the results of this study are confirmed and if they receive wide acceptance, major changes in the management of dairy cows will occur.

Implications of continued emphasis on high yield for cow health, reproduction and longevity

Grohn, Eicker and Hertl (1995) examined the relationship between previous 305-day milk yield and disease in 8070 Holstein cows of second and later parity from within 25 herds in New York State. It was felt that because many disorders occur early in lactation, it was better to use the previous 305-day mature equivalent production. Cows that calved between June 1990 and November 1993 were used in this analysis. A separate statistical model was used to study the occurrence of each of 7 disorders including: retained placenta, metritis, ovarian cysts, milk fever, ketosis, abomasal displacement, and mastitis. Table 8.5 shows the incidence risk as a percentage and the median day of occurrence postpartum. Only mastitis showed an increased incidence with increasing milk yield. However, it was cautioned that this did not necessarily mean a cause and effect relationship. It was explained that often cows with mastitis and low production are culled, whereas higher producing cows with mastitis may be kept in the herd as treatment is applied. Therefore, the continued presence of higher yielding cows with mastitis in the herd may cause an apparent relationship even if none is present.

Although studies reported prior to 1975 showed little relationship of higher milk yields to reproductive performance, later work has accumulated a considerable volume of research to show that some antagonistic relationship exists between high milk production and reproduction (Nebel and McGilliard, 1993). As these workers have noted, recent studies have described a detrimental effect of high yields particularly through a delay in ovarian activity and by a lower conception rate. But it was emphasized that managerial actions can have a major effect which greatly minimize the effect of the high milk production. In this context, adverse effects of an extended negative energy balance (NEB) can be minimized by formulation strategems such as use of supplemental fat to reduce the interval and

Table 8.5 LACTATIONAL INCIDENCE RISKS AND THE MEDIAN DAYS TO THE POSTPARTUM OCCURRENCE OF DISORDERS IN HOLSTEIN COWS[a]

Disorder	Lactational incidence risk (%)	Median postpartum day of occurrence (day)
Retained placenta	7.4	1
Metritis	7.6	11
Ovarian cyst	9.1	97
Milk fever	1.6	1
Ketosis	4.6	8
Displaced Abomasum	6.3	11
Mastitis	9.7	59

[a] Grohn *et al.*, (1995); 8070 cows in New York State

degree of NEB. No doubt some of this antagonism is expressed as dairy producers try to achieve the dogma of an ideal calving interval (CI) of 12 to 13.5 months. As noted above, if through the use of BST and/or other management tools, persistency can be maintained at a higher level and a CI of 13 months is no longer sought, then this antagonism may diminish greatly or even disappear.

There is much interest and concern for the nutrition and management of the transition cow, defined as the peripartum period from about 21 days prepartum to 15 to 30 days postpartum. Some work suggests that using low potassium diets and/or adjusting the cation/anion relationship to near zero or slightly negative during the 2 to 3 week prepartum, will result in less subclinical as well as clinical hypocalcemia and less depression in feed intake at parturition which will result in reducing the magnitude of NEB in the early postpartum. If this strategem is successful, it will probably diminish the degree of antagonism between high yields and reproductive performance.

Heat stress and high yields

For those of us who live in the sub-tropics or tropics, another dimension to higher milk yields is that the greater the feed intake, the greater the metabolic heat production. As Dennis Armstrong says, "a cow is a little furnace". And the greater the feed intake, the hotter the furnace. In summer the modern dairy cow does not belong in the sun, so shades, sprinklers and fans are all important for the alleviation of heat stress. But the primary reason for a decline in milk yield in hot weather is a voluntary reduction by the cow in feed intake. From a nutritional perspective, Chandler (1994) states that some feed ingredients have a lower heat increment (the heat associated with the metabolism of nutrients) than others. These include the fats especially, and in general, the lower fibre ingredients. But apart from nutrition, it is clear that if dairy producers in the warmer regions of the world are to keep up with the pace of increasingly higher production, then technologies which reduce heat stress will become increasingly mandatory.

Problems of formulation

The Cornell Net Carbohydrate and Protein System (CNCPS) is our most advanced system of feed formulation, but it too is undergoing a nearly constant revision (Barry *et al.*, 1994). It uses both carbohydrate and protein fractions that are partitioned according to their ease and speed of degradation in the rumen. But nearly all formulation programs require definitions of degradable intake protein (DIP) and undegradable intake protein (UDP). The National Research Council (NRC) for dairy cattle (1989) recommends 35% UIP and 60% DIP for a cow producing 40 kg of milk per day. Recently, Huber and Santos (1996) summarized the responses of a number of research studies from the literature which compared diets in which soyabean meal protein was replaced with a less degradable protein such as blood meal, brewers grains, feather meal, fish meal and blends of these. From 97 comparisons which involved lactation trials published from 1985 to 1994, it was found that milk yield increased in only 19% of the comparisons, there was no significant change in 73% and there was a significant decline in 9% of the trials. And if better protein nutrition was the goal in these studies, it was not reflected in the milk protein percentage because there was no change or even decreases in most of the trials. This comparison shows that much more care and refinement is needed in the successful application of the UIP/DIP system. Because microbial protein produced in the rumen has the best amino acid profile for milk synthesis, much more effort should be directed to those conditions which maximize the growth of ruminal microbes. In addition, the UIP protein needs an amino acid profile which complements the ruminal microbes. In the above comparison, fish meal substituted for soyabean meal resulted in greater milk yields in 46% of the trials. A major problem for feed formulators is that there is no well accepted method for the determination of DIP/UIP. Therefore, too often we are faced with using best estimates based on book values or from laboratory methods which are at best, compromises.

To provide optimal substrate for ruminal microbes in formulation, an expression for non-fibre carbohydrates (NFC) is needed. Again, lack of rapid, inexpensive laboratory procedures presents a serious obstacle. As Hoover and Miller (1996) note, the determination of NFC, also called nonstructural carbohydrate (NSC) by the difference method (easier method used frequently by feed testing laboratories compared to the enzymatic method which is more tedious), shows large differences for some feedstuffs. The objective is to measure or designate the sugars and starches, the rapidly digestible carbohydrates, but until more data become available from the enzymatic method, or another more rapid determination becomes available, progress is seriously hampered.

Another formulation quandary relates to the appropriate energy expression to use at the production intakes of lactating cows. The problem arises because there is a substantial disagreement between two authorities, the NRC (1989) for dairy cattle and Van Soest, Rymph and Fox (1992), especially for certain co-products. The disagreement arises over the amount of the depression in digestibility which occurs as intakes go from maintenance to 3 or 4 times maintenance. This depression occurs as cows eat more feed which results in a faster rate of passage through the digestive tract and a hence a lower digestibility for the feed. But how much less? The NRC assumes 8% less for all feeds at an intake of 3 times the intake equal to that required for maintenance. Van Soest *et al.* (1992) say that each feed ingredient differs, based on its own special characteristics. In Table 8.6 a comparison of the 2 systems is shown for several co-products. Some large differences are obvious. So what is a dairy producer or a ration formulator to do in light of this controversey? For the time being NRC (1989) is recommended, but it must be remembered that the energy values attached to some of the co-products may not truly reflect their real energy values, especially under some conditions. When a substitution is made to include a certain commodity, do high producers at their peak maintain their production well or even increase, or do they decline? Although this is very subjective, it may be the best indication we have that an appropriate energy value has been used for that co-product.

Conclusions

An old saying seems as true today as it was 50 years ago: "cows are better bred than fed". Geneticists have done an incredible job with their science; but this is good news to nutritionists and those in the feed business. Now when ration changes are made which are true improvements, cows usually respond with greater milk yields.

The food industry generates a large array of byproducts which ruminants can use productively to produce milk and meat. This is not to suggest that there are a lot of free lunches out there, because in a free market feed ingredients tend to be sold at a price which reflects their true nutritive value as buyers see them. Yet the availability of these products extends the feed supply and in some cases make special contributions to the diet.

Table 8.6 COMPARISON OF NET ENERGY FOR LACTATION BY VAN SOEST *et al.* (1992) AND NRC FOR DAIRY CATTLE (1989)

Feed	Discount[a] (%)	(NE_l)	
		3M[a]	NRC[b]
		— (MJ/kg DM) —	
Molasses-Cane	0.0	6.19	6.86
Bakery Waste	2.0	9.00	8.62
Alfalfa Hay	3.2	6.57	6.30
Whole Cottonseed	4.0	9.20	9.33
Soyabean Meal (44%)	5.1	7.45	8.17
Rice Bran	6.6	6.49	6.69
Wheat middlings	7.0	7.95	6.57
Brewers Grains (wet)	10.7	5.61	6.28
Corn Distillers (wet)	14.0	6.90	8.33
Corn Hominy	15.0	6.74	8.41
Pineapple Bran	18.0	4.35	6.49
Soya Hulls	18.0	5.40	7.41

[a] Van Soest *et al.* (1992);
[b] NRC (1989), discounted by 8% for 3M (see text for details).

About the time that dairy producers began to expand their herds, and mixer wagons with electronic load cells became available, the TMR system of feeding evolved and now is used widely. At first it was used before its inherent advantages were recognized, but today a large majority of enthusiastic users attest to its value. In larger herds, cows are housed in groups for managerial reasons, so that different formulations of TMRs can be used without special housing.

A number of dairymen have tried 3X milking but have returned to 2X milking. The primary reason seems to be that the initial increase in milk seen did not seem to persist, or cows were required to stand on concrete for an extended period and feet and leg problems developed. But an increase in milk yield must be paid for at some point with additional feed nutrients, and the opportunity for cows to eat that extra feed may not have been there in some cases.

The approval of BST for commercial use in the U.S. was accompanied by some controversey, despite years of research done in quadruplicate. Most of the controversey has disappeared or at least it has quietened down. The consistency of response of cows to this product and its successful use by some of the best herds has not yet caused its widespread adoption, despite a clear economic payback. This will probably come with time.

Despite a rather general impression that high yields are accompanied by increased incidence of metabolic problems, except for mastitis, this does seem to occur. There is some antagonism between reproductive competence and high yields, but most dairy producers seem willing to accept this downside. If current research comparing 13.5 month calving intervals with 18 months in conjunction with BST use is supported by additional work, even the lower reproductive performance may diminish or disappear.

Hot weather is a major constraint during summer in many parts of the U.S. If herd owners in the more humid and hotter areas of the world are to keep pace with their colleagues in more temperate regions, major expenditures must be made on technologies for heat stress abatement.

Feed formulators could be much more precise in the construction of diets if more precise values could be used for DIP/UIP, NSC, and net energy. Today, much uncertainty exists for some of the numbers which we use for these nutrient expressions. The bottom line is that much excess supplementation results and considerable money is wasted. Despite these uncertainties, production continues upwards with no limit in sight.

References

Albright, J.L. (1983) Putting together the facility, the worker and the cow. *Proceedings 2nd National Dairy Housing Conference*, 15–22.

Anonymous (1996) *Part I. Reference of 1996 Dairy Management Practices.* National Animal Health Monitoring System.APHIS-VS, U. S. Dept. Agriculture.

Barry, M.C., Fox, D.G., Tylutki, T.P., Pell, A.N., O'Connor,J.D., Sniffen, C.J. and Chalupa, W. (1994) *A manual for using the Cornell Net Carbohydrate and Protein System for evaluating cattle diets.* Revised for CNCPS Release 3, Sept. 1, Cornell University 14853–4801.

Bath, D.L., Dunbar, J.R., King, J.M., Berry, S.L., Leonard, R.O. and Olbrich, S.E. (1980) Byproducts and unusual feedstuffs in livestock rations. *W. Regional Ext. Publ.* 39.

Butcher, K.R. (1996) Personal Communication.

Chandler, P. (1994) Is heat increment of feeds an asset or liability to milk production? *Feedstuffs*, **66**, No. 15 12–17.

Chase, L.E. (1993) Developing nutrition programs for high producing dairy herds. *Journal of Dairy Science*, **76**, 3287–3293.

Coppock, C.E., Bath, D.L. and Harris, Jr., B. (1981) From feeding to feeding systems. *Journal of Dairy Science*, **64**, 1230–1249.

DePeters, E.J., Smith N.E. and Acedo-Rico, J. (1985) Three or two times daily milking of older cows and first lactation cows for entire lactation. *Journal of Dairy Science*, **68**, 123–132.

Gisi, D.D., DePeters, E.J. and Pelissier, C.L. (1986) Three times daily milking of cows in California dairy herds. *Journal of Dairy Science*, **69**, 863–868.

Grasser, L.A., Fadel, J.G., Garnett, I. and DePeters, E.J. (1995) Quantity and economic importance of nine selected by-products used in California dairy rations. *Journal of Dairy Science*, **78**, 962–971.

Grohn, Y.T., Eicker, S.W. and Hertl, J.A. (1995) The association between previous 305-day milk yield and disease in New York State dairy cows. *Journal of Dairy Science*, **78**, 1693–1702.

Gunderson, S. (1992) How six top Wisconsin herds are fed. *Hoard's Dairyman*, **137**, 686–687.

Halladay, D. (1996) Small wonders. *The Western Dairyman*, **77**, (No.9) 8–12.

Hoover, W.H. and Miller, T.K. (1996) Feeding for maximum rumen function. *Proceedings of Mid-South Ruminant Nutrition Conference*, 33–46. Texas Animal Nutrition Council, Dallas, TX

Huber, J.T. and Santos, F.P. (1996) The role of bypass protein in diets for high producing cows. *Proceedings of Southwest Nutrition and Management Conference*, 55–65. U. Arizona.

Jordan, E.R. and Fourdraine, R.H. (1993) Characterization of the management practices of the top milk producing herds in the country. *Journal of Dairy Science*, **76**, 3247–3256.

Merrill, L.S. (1996) 24,000 pounds of milk...no corn silage, no alfalfa. *Hoard's Dairyman*, **141**, 436.

Mowery, A. and Spain, J.N. (1996) Results of a nationwide survey of feedstuffs fed to lactating dairy cattle. *Journal of Dairy Science*, **79**, (Suppl. 1) 202. (Abstr.)

Muller, L.D. (1992) Feeding management strategies. In *Large Dairy Herd Management - 1992*, pp 326–335. Edited by H.H. Van Horn and C. J. Wilcox. Management Services, ADSA, Champaign, IL.

National Research Council (1989) *Nutrient Requirements of Dairy Cattle. 6th Revised Edition, National Academy of Science*, Washington, DC.

Nebel, R.L. and McGilliard, M.L. (1993). Interactions of high milk yield and reproductive performance in dairy cows. *Journal of Dairy Science*, **76**, 3257–3268.

Nizamani, A.H. and Berger, P.J. (1996) Estimates of genetic trend for yield traits of the registered Jersey population. *Journal of Dairy Science*, **79**, 487–494.

Spain, J.N. (1995) Management strategies for TMR feeding systems. *Proceedings 2nd Western Large Herd Management Conference*, 161–168. Las Vegas, NV.

Speicher, J.A., Tucker, H.A., Ashley, R.W., Stanisiewski, E. P., Boucher, J.R. and Sniffen, C.J. (1994) Production responses of cows to recombinantly derived bovine somatotrophin and to frequency of milking. *Journal of Dairy Science*, **77**, 2509–2517.

Tyrrell, H.F., Brown, A.C.G., Reynolds, P.J., Haaland, G.C., Peel, C.J. and Steinhour, W.D. (1988) Effect of Somatotropin on metabolism of lactating cows:energy and nitrogen utilization as determined by respiration calorimetry. *Journal of Nutrition*, **118**, 1024–1030.

Van Amburgh, M.E., Galton, D.M., Bauman, D.E. and Everett, R.W. (1997) Management and economics of extended calving intervals with use of BST. *Journal of Dairy Science* (in press).

Van Soest, P.J., Rymph, M.B. and Fox, D. (1992) Discounts for net energy and protein - fifth revision. *Proceedings Cornell Nutrition Conference* 40–68. Ithaca, NY.

Walton, R.E. (1983) A glimpse at dairying in the year 2000. *Dairy Science Handbook*, **15**, 511.

9

DEVELOPMENTS IN THE INRA FEEDING SYSTEMS FOR DAIRY COWS

R. VÉRITÉ[1], P. FAVERDIN[1] and J. AGABRIEL[2]
Institut National de la Recherche Agronomique,
[1] *Station de Recherche sur la Vache Laitière, INRA, 35590 Saint Gilles, France*
[2] *Laboratoire 'Adaptation des Herbivores au Milieu', INRA, 63122 Saint Genès-Champanelle, France*

In dairy production, optimal rationing is a most important key in controlling or improving animal performances (yield, reproduction, health, etc), milk quality, feed efficiency and N-wastage, and hence economical return. Modern feeding systems, as proposed in many countries, are useful tools for rationing. They are centred on better evaluation of nutritive values of feedstuffs and diets (mainly energy and protein) and of animal requirements or recommended allowances.

In France, the INRA feeding systems for ruminants have been elaborated from 1975 to 1977 (INRA, 1978; Vermorel, 1978; Vérité, Journet and Jarrige, 1979). They first included a net energy system (feed unit: UFL), and a metabolisable protein system (the PDI system e.g. protéines digestibles dans l'intestin). A system for predicting feed intake (the fill system) was developed slightly later (Jarrige *et al.,* 1979). They rapidly became fully functional.

They were revised ten years later (INRA, 1987; INRA, 1988; INRA, 1989), taking advantage of the information arising from research efforts as well as from their broad utilisation in practice. In addition at the same time, the above basic systems were co-ordinated in a powerful computer software package called 'INRAtion' that represents an integrated global system and allows easy rationing.

Since then several extra concepts have been developed (i.e. requirement for individual amino acids, marginal response curves) and several tools have been proposed to promote easier implementation. A new updating, based on the same scheme, is planned for the next 3-4 years. Simultaneous research projects aim at elaborating a nutrient-based model to better account for complex regulations and for differential animal responses.

As the current energy and protein systems are described in detail elsewhere (Vermorel, Coulon and Journet 1987; Vérité *et al*, 1987; INRA, 1989), only some new developments of the protein system will be presented here. The main purpose of this paper is rather to stress several other important topics that received particular attention in the INRA systems: feed intake estimation, responses to marginal input variations, integrated software as a practical tool for rationing (INRAtion).

Developments in the PDI system

The PDI system is similar to several other systems that were published later when considering the concepts and the frameworks, according to several recent comparisons; however they differ to some extent when considering available functionalities and real estimation of diets in practice (see Vérité and Peyraud, 1995 among others). With regard to the basis of the PDI system, attention should be drawn to the large experimental data set that was used for elaborating the basic relationships, the tables of feed values and the animal requirements. It should also be noted that most of the basic parameters and relationships were elaborated in a coordinated manner to prevent bias in rationing, as shown during the validation steps.

Protein degradability is basically assessed by the nylon bag method but this has some well known shortcomings: variability between laboratories, low accuracy and high cost. Therefore special attention was paid to careful standardisation, along with the use of the same reference sample in every series of measurements (Michalet-Doreau, Vérité and Chapoutot, 1987; Vérité *et al.*, 1990). Furthermore, as commercial laboratories require alternative methods, an enzymatic degradability test was developed for concentrate evaluation, standardised and then calibrated against the basic nylon bag procedure (Aufrère *et al.*, 1991). This is now implemented in French commercial laboratories. Similar work is in progress on roughages. With regard to animal requirements, much attention was paid to response curves of animal and rumen outputs to marginal N inputs (see later).

A great deal of work has been done on the concept of amino-acid profile in dairy diets. The initial interest came from a large set of INRA and foreign trials where extra supplies of lysine and methionine increased milk protein content and yield (Rulquin *et al.*, 1993). Therefore, the PDI system was expanded to include the concept of LysDI and MetDI, that evaluates the amounts of methionine and lysine digested in the small intestine (called respectively MetDI and LysDI) (Rulquin and Vérité, 1993). Feed values are derived from the PDI concepts with additional account of the amino-acid profiles of microbial and specific feed proteins. Recommended allowances are expressed as a percentage of PDI requirements, in accordance with the ideal protein concept. Requirements were found to be 2.5 and 7.3% PDI for MetDI and LysDI respectively. However values of 2.0 and 6.8% are considered as thresholds below which one could start considering supplementation (milk protein response vs. extra cost). This extension has been available for practical use since 1992 in advisory papers, and was included in the INRAtion software from 1995.

Predicting feed intake: the Fill Unit system

An accurate estimate of voluntary intake of roughages and/or total diet is the first and most important step when rationing because of significant variability with many factors, including rationing strategy itself. In general, direct estimates of dry matter intake rely on empirical relationships involving only a few simple characteristics of diets and animals. Such methods could work soundly in common situations, but could have important shortcomings in situations outside the range of initial data and are irrelevant to simulate the effect of different strategies (Faverdin, 1992; Faverdin, Baumont and Ingvartsen, 1995).

Several other methods are based on the widely accepted assumption that intake in ruminants is regulated through two main concomitant mechanisms involving physical and metabolic (or energy) limitations. In a first group of methods (Forbes, 1977; NRC, 1987; Mertens, 1987) both bulk and energy constraints are considered separately thus involving two independent parameters (and units) to characterise the satiating properties of the diet. For each constraint, the sum of the value of the feeds ingested cannot exceed the animal capacity; maximum intake is assessed from the first occurring limitation. In the other approach, relating to the French and Danish 'fill systems', both constraints are described with a unique parameter, the 'fill' value (FV), that accounts for all components of the satiation value of the feeds. They work like a nutritive system assuming only, with one common unit, that the diet 'fill value' equals the animal 'intake capacity'.

In the French fill system, the main basis for such a simplification relies on a relevant utilisation of the concept of substitution rate between concentrate and forage. The system works as if the satiating signals from the bulk and energy regulations interact with each other in a cumulative way, at least to some degree. Therefore the term 'satiation' would have been more appropriate than 'fill' to designate the system and its unit.

The fill system was outlined by Jarrige *et al.* (1979; 1986) and revised by Dulphy, Faverdin and Jarrige (1989). The aim was to develop a practical and additive system capable of predicting intake in cattle, sheep and goats under various circumstances. Most of the basic elements of the system were developed from several sets of feeding experiments run to determine 1) the quantity of feed that the animal is able to eat voluntarily according to its appetite in order to predict the feed intake capacity; 2) the voluntary dry matter intake (or ingestibility) of each forage when given *ad libitum* as the sole food to ruminants in order to estimate the 'fill value' of the forage' and 3) the influence of the supplementary concentrates upon the voluntary intake of forages, i.e. the substitution rate in order to calculate the 'fill value' of the concentrate. As a whole, the system appears to provide estimates of voluntary dry matter intake (VDMI) that fit reasonably well with observed data (Ostergaard, 1978) as illustrated in Figure 9.1.

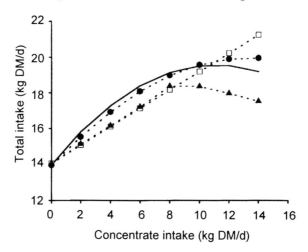

Figure 9.1 Effect of concentrate level on dry matter intake : comparison of experimental data (—) from Ostergaard (1978) with estimates from the NRC (▲···▲), the Danish Fill Unit (□···□) and the INRA Fill Unit (●···●) systems.

BASIC PRINCIPLES

A common unit, 'the fill unit' (FU) was first defined to describe both the intake capacity (IC) of animals and the satiating properties of feedstuffs i.e. their fill value (FV). By definition, one 'fill unit' corresponds to the fill value of one kg DM of a reference forage that is an average pasture grass with specific characteristics. It works just as for the energy system, where the net energy content of one kg reference barley is the feed unit (UFL).

If at least one feed at least is offered *ad libitum*, the weighted sum of the fill values of all the feeds is equal to the intake capacity of the animal. Therefore on *ad libitum* feeding, calculation of diet or forage intake results from the following equation:

$$IC = \Sigma \ Q_i \ x \ FV_i \qquad (1)$$

where IC is the intake capacity (FU/d), Q is feed intake levels (kg DM), FV is the feed fill value (FU/kg DM) and i refers to the different feeds in the diet.

Intake capacity is a fixed attribute of an animal in a given status whatever the diet characteristics. The fill value is a fixed attribute of a forage batch whatever the final diet and the animal characteristics. It is an inverse function of its ingestibility i.e. its VDMI when fed *ad libitum* as the sole feed to a standard animal: the higher is the ingestibility, the smaller is the fill value. The fill value (FU/kg DM) of a test forage is obtained relatively to the reference forage:

$$\text{test forage FV (FU/kg DM)} = \frac{\text{reference forage ingestibility}}{\text{test forage ingestibility}} \quad (2)$$

Therefore the forages FV value accounts mainly for both physical and palatability factors.

Unlike the two others parameters, the fill value of a concentrate is not a fixed attribute for that concentrate but depends also on the current nutritional situation. This accounts for metabolic regulation mechanisms as reflected by the observed variations in the substitution rate between forage and concentrate. Adding a concentrate to forage offered *ad libitum* usually increases the total VDMI but reduces the forage VDMI. As forage FV and intake capacity are independent of concentrate intake, the FV of concentrate is a function of the substitution rate between concentrate and forages. The fill value of the concentrate (FVc) is then derived from the equation:

$$FV_c = FV_f \times SR \quad (3)$$

where FV_f is the roughage fill value and SR is the substitution rate between roughage and concentrate. In other words, one kg extra concentrate has the same 'fill' effect as the amount of forage which has disappeared, because IC is constant for a given animal situation. Contrary to the FV of forages, the FV of a concentrate reflects mainly the satiating effect through metabolic regulation i.e. mainly energy regulation.

Many experiments have shown that the substitution rate is highly variable. With different types of forages and concentrates, it has been observed that the substitution rate increases with the energy balance of the animal (Faverdin *et al.,* 1991). From these data, a model was developed to predict that substitution rate increases with the energy balance from zero up to a plateau higher than 1 according a complex function (Dulphy *et al.,* 1987). However when estimating the substitution rate, the energy balance is not known *a priori* but would arise from the resulting estimation of diet intake. Therefore the calculations should start from forage FV and energy characteristics of concentrate, forage and requirements and develop in an iterative way. The complete calculation is available within the INRAtion software.

PRACTICAL USE

All the values for animal IC and forage FV are available from tables given in reference papers (Andrieu, Demarquilly and Sauvant, 1989) or in the INRAtion software. Forage FV values could also be estimated for each batch from proximate analysis or plant characteristics such as morphological traits or age (through the estimation of VDMI by Demarquilly, Andrieu and Weiss, 1981).

The values of IC varies with animal species, physiological stage, and productive traits. In dairy cows, IC is 17 FU/d for a 600 kg cow producing 25 kg FCM and increases with milk yield (0.1 FU/kg FCM) and liveweight (0.01 FU/kg), parity (10% lower in primiparous); values are lower in early lactation (coefficients for the first 12 weeks).

The tabulated FV values for forages (Andrieu et al, 1989) are derived from an initial database with VDMI measurements in sheep for more than 3000 forage samples. The conversion and calculation methods for estimating FV in cattle are given by Dulphy, Faverdin and Jarrige (1989). The FV values for forages range between 0.9 and 1.6 FU/kg DM. They vary with many characteristics such as the initial characteristics of plants (species, physiological stage, chemical composition), the type of conservation (hay or silage), the drying method (barn-dried or field cured), the weather conditions, the harvesting method (finely chopped or flail harvested), the use of additives, and the dry matter content of the silage.

The range of FV for a given concentrate is rather large (0 to possibly more than 1) as shown in Table 9.1. Individual concentrates are not ascribed a particular FV in tables as FVc depends mainly on the energy balance of the diet (cf. above). The INRAtion software does not require FV of concentrate as an input because of automatic calculation. However the FV of a concentrate is also available from simplified tables or equations starting from forage characteristics (FV and UFL) and milk yield.

Table 9.1 VARIATION OF CONCENTRATE FILL VALUE ACCORDING TO FORAGE QUALITY AND FAT-CORRECTED MILK (FCM) YIELD FOR LACTATING DAIRY COWS FED WITH 5kg DM OF CONCENTRATES

Forage characteristics		Fill value of concentrate	
Fill value (FU)	*Energy (UFL)*	*20 kg FCM*	*35 kg FCM*
1.3	0.75	0.37	0.15
1.1	0.75	0.49	0.19
1.2	0.90	0.64	0.20
1.0	0.90	0.86	0.30

Productive responses to marginal variations in feed supply

In most modern systems, rationing starts with the desired level of animal output and works back to calculate the feed inputs required to sustain the corresponding assumed requirements (Gill, 1996). However in practice, the reverse way is of similar if not greater interest to the farmers. Most often the problem is not to elaborate an entirely new diet but rather to change some particular feed ingredient(s) in the current diet. This could arise because of problems with feedstuff availability, performance or health of animals, or economical changes (interest in deviating from the recommended allowances). Then it becomes important to know the relationships between any marginal changes in the diet and the induced changes in animal performance and efficiency.

The marginal responses could differ greatly from the theoretical ones that would be expected from average efficiency values. Productive traits usually respond according the law of diminishing return when nutritive supplies vary around requirement levels. At the moment, such input-output relationships are not easy to assess from theoretical approaches. However they could be developed from feeding trials, each comparing different input levels of a sole nutritive parameter (energy, PDI, etc).

Such response curves of intake and milk or protein yields to changes in the levels of concentrate, or energy, or metabolisable protein, or degradable protein, or specific amino acids (lysine and methionine) were obtained from relevant INRA and foreign data. Then they were either incorporated in or added as extra tools to, the basic 'energy', 'protein' and 'fill' systems. Therefore, such effects are easily simulated or anticipated with the INRAtion software.

For instance, it is generally accepted that the marginal efficiency of concentrate for milk production (extra milk/extra concentrate) decreases as concentrate allocation increases around the 'required level'. This arises from changes 1) in voluntary DM intake because of substitution, 2) in efficiency of digestive and metabolic processes and 3) in partition of available nutrients between different tissues. The INRA energy and fill systems account for such mechanisms. Firstly, the expected changes in feed intake are directly simulated by the fill system as shown in Figures 9.2 and 9.3. For a given cow, increasing concentrate allocation emphasises the negative effect on forage intake of an extra kg of concentrate and diminishes the positive effect on total DM intake. Further, when starting from a similar concentrate level, these effects are greater 1) when the yield of milk is lower (Figure 9.2) or 2) when the quality of the forage is better i.e. when forage energy concentration and ingestibility are higher as for instance, with good quality maize silage compared to medium quality grass silage (Figure 9.3). Secondly, the energy system (FU) accounts directly for a reduced digestive/metabolic efficiency when the feeding level and/or the concentrate:forage ratio are increased. These

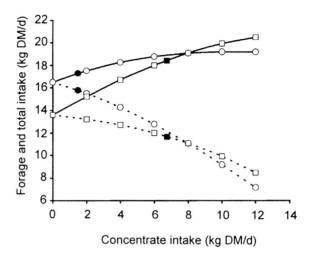

Figure 9.2 Levels of forage (---) and diet intake (—) according to the amount of concentrate supplied and the nature of forage (good quality maize silage O vs medium quality grass silage □) as simulated with INRAtion for a 25 kg FCM cow (full symbols ■, ● indicate the situations when requirements are met).

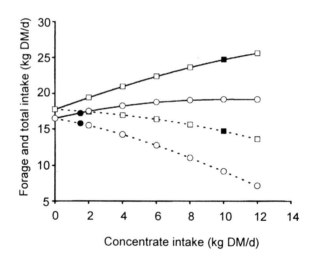

Figure 9.3 Levels of forage (---) and diet intake (—) according to the amount of concentrate supplied and milk yield level in mid lactation (25 kg O vs 40 kg □ of FCM) as simulated with INRAtion on a maize silage diet (full symbols ■, ● indicate the situations when requirements are met).

effects could account for up to 10% of the total net energy theoretically supplied (INRA, 1987). Lastly, the response curve of milk yield to changes in the real net energy availability was empirically derived from feeding trials through regression analysis (Faverdin, Hoden and Coulon, 1987). Therefore, the expected effects on animal performance (milk and protein yields and body-weight change) are directly and automatically provided, as shown in Table 9.2. In that simulation, an extra kg concentrate over requirement level would increase milk yield by only 0.5 kg FCM. The decreases in forage intake and in digestive efficiency would have accounted for approximately 45 and 10% of the net energy theoretically supplied by the extra concentrate; the remaining 45% really available would have been split between milk (2 parts) and body weight gain (1 part). If the initial situation corresponds to underfeeding, the calculated milk response is higher and the other effects are lower.

Table 9.2 MARGINAL EFFECTS OF AN EXTRA KG OF CONCENTRATE AS SIMULATED BY THE INRAtion SOFTWARE FOR A 600 kg DAIRY COW PRODUCING 30 kg FCM, INITIALLY FED AT REQUIREMENT WITH 5.4 kg DM CONCENTRATE AND MAIZE SILAGE *AD LIBITUM* (15.1 kg DM)

	Marginal response	*Net energy partition (%)*
Concentrate intake	+1 kg as fed	(100)
Maize silage intake	-0.5 kg DM	50
Milk yield	+0.5 kg FCM	25
Milk protein content	+0.3 g/kg	
Body-weight change	0.05 kg/d	15
(Lower efficiency)		10

Similarly, the marginal increase in milk yield in response to a same amount of extra metabolisable protein (PDI), is lower still since the initial PDI supply, as a proportion of requirement, is higher. Data from several INRA experiments with grass and maize silage diets, each comparing several PDI levels at constant DM intake, are summarised in Figure 9.4 (Vérité, Dulphy and Journet, 1982; Vérité, unpublished). It appears that, after peak yield, the marginal responses to PDI variations is close to 0.5 kg FCM/100 g extra PDI when total input is close to requirement level. It reaches 0.9 kg FCM/100 g extra PDI if the initial PDI balance is negative by 200 g and only 0.2 kg FCM/100 g extra PDI if the initial PDI balance is positive by 200 g. These relationships could help in deciding if protein supply should meet requirement or deviate from it in order to optimise economical return according to the costs of energy and protein sources and to milk price. The effects are lower in early lactation because protein mobilisation could compensate to some degree. During the first 6 to 8 weeks postcalving, the cumulative negative PDI balance could reach 5 to 10 kg PDI, without significant negative effect on milk yield. Similarly, marginal response of protein content and yield to variations in lysine and methionine supply are now available (Rulquin *et al.*, 1993).

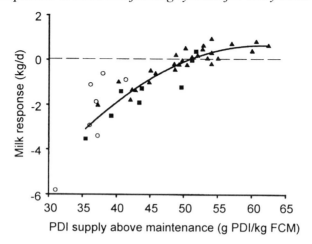

Figure 9.4 Response of milk yield to the level of PDI supply in dairy cows fed maize or grass silage diets in early (O) peak (■) or mid (▲) lactation (the regression line does not include early lactation data). Milk responses are calculated within experiments by reference to the control group fed at maintenance, i.e. 48 g PDI/kg FCM.

A shortage in the degradable protein supply is generally assumed to negatively affect microbial activity in the rumen and animal performance. Nitrogen recycling in the digestive tract could reduce that effect to some degree. Only few systems account for such a possibility, most often in a fixed way. However it could be modulated by other factors as shown by several INRA feeding trials, each comparing different levels of degradable protein at similar dry matter and metabolisable protein intakes (Figure 9.5). The negative effect of degradable protein shortage would depend on the animal balance for metabolisable protein. The adverse effect on milk yield of a negative 'rumen balance' (i.e. degradable protein supply as a proportion of microbial requirement) is higher when the 'animal balance' (i.e. metabolisable protein supply as a proportion of animal requirement) is lower. Further, when the 'animal balance' is positive, the rumen could afford a significant deficit in degradable protein supply. This effect is most probably related to a greater potential for nitrogen recycling because then the excess amount of amino acids is catabolised into urea. Roughly, the degradable protein supply should provide at least 100, 92 or 85% of microbial requirement if the animal balance (PDIE minus animal requirement) is respectively negative, zero or positive. Therefore, with the usual PDI parameters, the corresponding recommendation is that the PDIN/PDIE ratio in dairy diets should be at least 100, 96 or 92%.

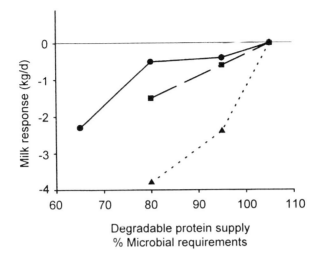

Figure 9.5 Variations of milk response to deficit in degradable protein supply according to animal balance in metabolisable protein (negative ▲···▲, normal ■--■ or positive ●—● balance).

The 'INRAtion' rationing software

The 'INRAtion' rationing software (INRA, 1995) has been available since 1987. It was designed as an aid to advisors and also to everyone involved in nutrition and feeding problems in cattle, sheep or goats under various physiological and productive situations. It integrates in a functional package, the intake constraint along with the nutritive constraints (energy, protein). It includes a fairly large data base on feedstuffs values (common concentrate ingredients, 600 forages from temperate areas and also 250 forages from Mediterranean, and humid or dry-tropical areas). It can also be modified by each user with his own feedstuffs analyses. Different feeding strategies can be explored, covering or deviating from requirements. Such flexibility gives a high degree of adaptation to every particular local or economical constraints. Similarly, it manages individual rationing as well as with totally mixed diets.

When starting with a given production level as a set point, it allows 'optimisation' of the diet or concentrate in terms of ingredients composition and/ or amount to be supplied (possibly giving several solutions), while voluntary DM intake of forages is also predicted. The diets are directly and automatically optimised on intake, energy and protein constraints, while further optimisation on LysDI and MetDI, minerals, trace elements, vitamins can be performed when desired.

In the reverse way, when starting from a given diet, different dietary changes can be proposed and are tested for their effects on ingestion and nutritional parameters while the resulting marginal effects on production parameters are assessed from the response curves mentioned above.

Beyond its value as an aid to technical and economic problems, such software is also of good value for educational purposes, favouring the spread of common concepts and references for dairy farmers, students, advisors and feed industry. Most of the teaching, advisory or milking control services, and feed manufacturers have access to INRAtion. Furthermore, in several situations it has been included as a basic part of other software packages that were developed for broader purposes.

References

Andrieu, J., Demarquilly, C. and Sauvant, D. (1989) Tables of feeds used in France. *In Ruminant nutrition*, pp 213–303. Edited by R. Jarrige. John Libbey: London, Paris.

Aufrère, J., Graviou, D., Demarquilly, C., Vérité, R., Michalet-Doreau, B. and Chapoutot, P. (1991) Predicting in situ degradability of feed proteins in the rumen by two laboratory methods (solubility and enzymatic degradation). *Animal Feed Science and Technology*, 33, 97–116.

Demarquilly, C., Andrieu, J. and Weiss, P. (1981) L'ingestibilité des fourrages verts et des foins et sa prévision. In *Prévision de la valeur nutritive des aliments des ruminants,* pp 155–168. Edited by C. Demarquilly. INRA Publications: Versailles.

Dulphy, J.P., Faverdin, P. and Jarrige, R. (1989) Feed Intake: the Fill Unit Systems. In *Ruminant Nutrition: recommended allowances and feed tables,* pp 61–72. Edited by R. Jarrige. INRA, John Libbey Eurotext: Paris.

Dulphy, J.P., Faverdin, P., Micol, D. and Bocquier, F. (1987) Révision du système des unités d'encombrement (UE). *Bulletin Technique du Centre de Recherches Zootechniques et Vétérinaires de Theix*, 70, 35–48.

Faverdin, P., Baumont, R. and Ingvartsen, K.L. (1995) Control and prediction of feed intake in ruminants. In *Recent developments in the nutrition of herbivores. Proceeding of the IV International Symposium on the Nutrition of Herbivores,* pp 95–120. Edited by M. Journet, E. Grenet, M.H. Farce, M. Theriez and C. Demarquilly. INRA Editions: Versailles.

Faverdin, P., Dulphy, J.P., Coulon, J.B., Garel, J.P., Rouel, J. and Marquis, B. (1991) Substitution of roughage by concentrates for dairy cows. *Livestock Production Science*, 27, 137–156.

Faverdin, P. (1992) Les méthodes de prédiction des quantités ingérées. *INRA Productions Animales*, 5, 271–282.

Faverdin, P., Hoden, A. and Coulon, J.B. (1987) Recommandations alimentaires pour les vaches laitières. *Bulletin Technique C. R. V. Z. Theix*, **70**, 133–152.

Forbes, J.M. (1977) Development of a model of voluntary food intake and energy balance in lactating cows. *Animal Production*, **24**, 203–224.

Gill, M. (1996) Modelling nutrient supply and utilization by ruminants. In *Recent developments in ruminant nutrition (3)*, pp 23–34. Edited by P.C. Garnsworthy and D.J.A. Cole. Nottingham University Press: Nottingham.

INRA (1978) *Alimentation des Ruminants*. Edited by R. Jarrige. INRA Publications: Versailles.

INRA (1987) Alimentation des Ruminants: révision des systèmes et des tables de l'INRA. *Bulletin Technique CRZV Theix, INRA*, **70**, 1–222.

INRA (1988) *Alimentation des bovins, ovins et caprins*. Edited by R. Jarrige. INRA Publications: Versailles.

INRA (1989) *Ruminant nutrition. Recommended allowances and feed tables*. Edited by R. Jarrige. John Libbey: London - Paris.

INRA (1995) *INRAtion, logiciel d'aide au rationnement des ruminants. Version 2.60*. INRA-CNERTA: Dijon.

Jarrige, R., Demarquilly, C., Dulphy, J.P., Hoden, A., Robelin, J., Beranger, C., Geay, Y., Journet, M., Malterre, C., Micol, D. and Petit, M. (1979) Le systeme des unités d'encombrement pour les bovins. *Bulletin Technique C. R. V. Z. Theix*, 38, 57–79.

Jarrige, R., Demarquilly, C., Dulphy, J.P., Hoden, A., Robelin, J., Beranger, C., Geay, Y., Journet, M., Malterre, C., Micol, D. and Petit, M. (1986) The I.N.R.A. 'fill unit system' for predicting the voluntary intake of forage-based diets in ruminants: a review. *Journal of Animal Science*, **63**, 1737–1758.

Mertens, D.R. (1987) Prediction intake and digestibility using mathematical models of ruminal function. *Journal of Animal Science*, **64**, 1548–1558.

Michalet-Doreau, B., Vérité, R. and Chapoutot, P. (1987) Méthodologie de la mesure de la dégradabilité in sacco de l'azote des aliments. *Bulletin Technique CRZV Theix, INRA*, **69**, 5–7.

NRC (1987) *Predicting feed intake of food-producing animals*. National Academic Press. Washington.

Ostergaard, V. (1978) *Strategie for concentrate feeding to attain optimum feeding level in high yielding dairy cows*. National institute of animal science. Copenhagen.

Rulquin, H. and Vérité, R. (1993) Amino acid nutrition of dairy cows: Productive effects and animal requirements. In *Recent Advances in Animal Nutrition - 1993*, pp 55–77. Edited by P.C. Garnsworthy. Nottingham University Press: Nottingham.

Rulquin, H., Pisulewski, P.M., Vérité, R. and Guinard, J. (1993) Milk production and composition as a function of postruminal lysine and methionine supply: a nutrient-response approach. *Livestock Production Science*, **37**, 69–90.

Vérité, R., Dulphy, J.P. and Journet, M. (1982) Protein supplementation of silage diets for dairy cows. In *Forage protein conservation and utilization*, pp 175–190. EEC Seminar, Dublin.

Vérité, R., Journet, M. and Jarrige, R. (1979) A new system for the protein feeding of ruminants: the PDI system. *Livestock Production Science*, **6**, 349–367.

Vérité, R., Michalet-Doreau, B., Chapoutot, P., Peyraud, J.L. and Poncet, C. (1987) Révision du système des Protéines Digestibles dans l'Intestin (PDI). *Bulletin Technique CRZV Theix, INRA*, **70**, 19–34.

Vérité, R., Michalet-Doreau, B., Vedeau, F. and Chapoutot, P. (1990) Dégradabilité en sachets des matières azotées des aliments concentrés: standardisation de la méthode et variabilités intra et interlaboratoires. *Reproduction Nutrition Développement*, **Suppl 2**, 161–162.

Vérité, R. and Peyraud, J.L. (1995) Comparison of the modern protein evaluation systems for ruminants. In *Proceedings of the 3rd International Feed Production Conference, Piacenza (Italie) - feb 1994*, pp 233–247. Edited by G. Piva. Piacenza.

Vermorel, M. (1978) Feed evaluation for ruminants. 2. The new energy systems proposed in France. *Livestock Production Science*, **5**, 347–365.

Vermorel, M., Coulon, J.B. and Journet, M. (1987) Révision du système des Unités Fourragères (UF). *Bulletin Technique CRZV Theix, INRA*, **70**, 9–18.

10

DEVELOPMENTS IN AMINO ACID NUTRITION OF DAIRY COWS

B. K. SLOAN
Rhône-Poulenc Animal Nutrition, 42 avenue Aristide Briand, 92164 Antony, France

What are the potential benefits to balancing dairy rations for their digestible amino acid content?

The increasing pressure in the dairy industry to produce milk more efficiently while tailoring its composition to meet the demands of different market segments has given renewed impetus to devising appropriate feed formulations and feeding strategies to meet these challenges. Optimising the amino acid balance of rations has been proposed as one approach, to enhance milk protein secretion and manipulate milk protein content. However, to be widely adopted in practice a clear definition is needed of how to balance amino acids, appropriate feed formulation constraints need to be developed and the economic benefits demonstrated.

AN ALREADY PROVEN CONCEPT IN MONOGASTRIC NUTRITION

Optimising the supplies of individual amino acids has been common practice in poultry nutrition for over 30 years and in pig nutrition for the last 20 years. The principal advantage of doing this is an increase in animal performance - lean growth in broilers and growing pigs and egg production in laying hens, with an accompanying improvement in feed conversion efficiency. A similar improvement in performance may also be obtainable for ruminants by paying more attention to individual amino acid supplies, rather than trying to satisfy total amino acid supplies on a global basis. However, due to the role played by the rumen in transforming a large proportion of dietary protein into microbial protein, it has been very difficult to determine whether there is a significant performance advantage to increasing the supplies of specific individual amino acids post-ruminally for the dairy cow of high genetic merit.

In this chapter, the evidence of limitations to dairy cow performance arising from the quantity and relative proportions of individual amino acids absorbed from the small intestine of cows fed conventional rations will be reviewed. How this evidence has been transformed into a practical set of recommendations for digestible lysine and methionine concentrations in dairy rations has been outlined in the previous chapter (Vérité, 1997). These recommendations will be used as the basis for discussing the potential benefits of formulating rations to contain specific digestible lysine and methionine levels on various practical aspects of dairy cow performance.

DEFINITION OF THE AMINO ACID BALANCE CONCEPT FOR RUMINANTS

It should be stated from the outset, that when we talk about amino acid nutrition of ruminants, there is firstly a very important quantitative component and secondly a qualitative component that should not be neglected. Both are important in order to maximise performance and should be treated as a whole and not as alternatives. First and foremost, a ruminant needs a sufficient supply of all amino acids, essential and non essential, at the systemic level, to satisfy the large majority of amino acid demands for maintenance and production. This is normally achieved by formulating rations to satisfy the recommendations laid down in any modern protein system such as the metabolisable protein (MP) system (AFRC,1992) or the protein digested in the intestine (PDI) system described in the previous chapter. Following the principles of monogastric nutrition, a further gain in performance can be envisaged if the correct 'balance' or 'profile' of digestible amino acids is supplied. The improvement in performance in monogastrics is achieved by increasing the efficiency of total amino acid utilisation and reducing the energetic load for removing surplus amino acid nitrogen (Henry, 1993 ; Leclercq, 1996).

The qualitative component of amino acid nutrition in dairy cows will be regarded in most detail in this review. Specific reference to quantitative supplies will only be included where it is believed to be important to the understanding of certain responses observed.

Determining the limiting amino acids for milk performance

RATION PROTEIN SOURCE DETERMINES THE QUANTITY OF MILK PROTEIN SECRETED

Rulquin and Vérité (1993), from a compilation of trials where treatment differences in crude protein content and N-degradability were negligible, showed that the protein source in the diet could have a marked influence on milk protein secretion. Soyabean meal was superior to maize gluten meal and fishmeal was superior to soyabean. In fact, fishmeal-containing diets outperformed maize gluten meal rations by as much as 150 g of milk protein / cow / day. It would appear that the potential advantage of low protein degradability of maize gluten meal is compromised by its poor amino acid profile. The amino acid profile of the different protein sources would appear to have an important effect in determining the quantity of milk protein secreted.

Casein has been the mostly widely used protein source to investigate the general interest in 'improving' the quality of protein available to the dairy cow. Consistently large responses to casein infusions have been observed in terms of milk yield (1 to 4 kg per cow/day) and milk protein yield (50 to 180 g per cow/day) (Clark, 1975). There is apparently both an important quantitative as well as qualitative component to the initiation of this response. A large part of the volume response may be attributable to an increase in the total quantity of amino acids absorbed (PDI) by the dairy cow. It is also evident that the responses are very much related to the basal level of protein supply. When protein supplies are lower than requirements, a larger proportion of the additional available amino acid-N is partitioned towards milk protein synthesis relative to other productive functions compared with when requirements are already satisfied (Whitelaw, Milne, Ørskov and Smith, 1986).

However, the qualitative advantage of casein, due to its balanced amino acid profile, is also important, as demonstrated in trials comparing isonitrogenous quantities of casein and soya isolate (Choung and Chamberlain, 1993 ; Rulquin, 1986). Whereas 50 % of the extra amino acid-N from casein was found in milk protein, only 20 % was recovered as milk protein when soya isolate was infused.

NON ESSENTIAL AMINO ACIDS ARE GENERALLY NOT LIMITING

The infusion of casein has been mimicked by infusion of a solution of all its constituent individual amino acids in proportion. Improvements in milk protein secretion were essentially of the same magnitude as for casein and likewise when only the 10 so called essential amino acids were infused (Fraser, Ørskov, Whitelaw and Franklin, 1991; Schwab, Satter and Clay, 1976). This indicates that when supplies approach conventional recommendations for total amino acids (PDI or MP) there is no need for a marginal increase in non essential amino acid supplies; they are already supplied in excess and/or the animal can easily synthesize any additional requirements from other amino acids present in excess quantities.

METHIONINE AND LYSINE, THE FIRST TWO LIMITING AMINO ACIDS

Many trials have tried to identify the sequence of limiting amino acids for milk protein production by infusing individual amino acids post-ruminally. Schwab, *et al* (1976) demonstrated that methionine and lysine were likely to be the two most limiting amino acids in dairy rations for milk protein secretion. Increasing supplies of these two amino acids alone improved milk protein content by 1g/kg and milk protein secretion by 45 g/day, which accounted for 43 % of the response to an infusion of casein.

Many experiments of a similar nature have since been carried out over a wide range of diets with the focus of the investigations being the response to increasing digestible lysine (lysDI) and methionine (metDI) supplies together. Rulquin (1992) summarized much of this work   (Table 10.1) and showed quite clearly that in the vast majority of rations, lysine and methionine must be the first two limiting amino acids. By increasing these two amino acids alone, milk protein was increased on average by 35 and 50 g/head/day in mid and early lactation cows respectively.

Table 10.1 EFFECTS OF THE STAGE OF LACTATION ON THE RESPONSES TO A POST-RUMINAL SUPPLY OF METHIONINE AND LYSINE

Stage of lactation	Early lactation	Mid lactation	SED
Week	1 to 9	10 to 29	
No. of trials	(16)	(71)	
Amino acids supplies (g/day)			
Methionine	10	8	
Lysine	24	20	
Productions			
Milk (kg/d)	+0.7	+0.1	0.9
Protein (g/d)	+56	+31	
Fat (g/d)	+10	+1	51
Milk composition			
Protein level (g/d)	+1.2	+1.0	0.7
Fat level (g/kg)	-0.5	-0.0	1.6

Courtesy of Rulquin 1992

The relative importance of lysine limitations with respect to methionine was not so clear. Subsequent experimentation showed that for rations where maize protein was the major constituent of bypass protein supply, lysine was clearly first limiting (Rulquin, Lehenaff and Vérité, 1990; Robert, Sloan and Nozière, 1996; King, Bergen, Sniffen, Grant, Grieve, King and Ames, 1991). Protein secretion was increased by 50 to 150 g/cow/day where only lysine was increased post-ruminally. However, where maize protein made only a minor contribution and soyabean meal the major contribution to bypass protein, methionine is clearly the first limiting amino acid (Robert, Sloan and Lahaye, 1995; Rulquin and Delaby, 1994a ; Robert, Sloan and Denis, 1996; Armentano Bertics and Ducharme, 1993). In fact, supplying additional digestible lysine did not improve the milk protein response further (Sloan, 1993) on this type of diet. Where maize was the only cereal energy source and soyabean the principal protein source, Koch, Whitehouse, Garthwaite, Wasserstrom and Schwab (1996) showed methionine still to be clearly first limiting for milk protein synthesis, but further performance improvements could be obtained by providing extra digestible lysine.

IMPROVEMENTS IN MILK PROTEIN ARE A CONSEQUENCE OF AN INCREASE IN MILK CASEIN

The consistent increases in milk protein secretion and particularily in milk protein content observed (Rulquin, 1992) as a result of improvements in metDI and lysDI supply, are due to an increase in milk casein since the proportion of casein within milk crude protein or milk true protein is increased (Pisulewski, Rulquin, Peyraud and Vérité, 1996). Hurtaud, Rulquin and Vérité (1995) tested the cheese-making properties of milk from a range of trials where metDI and lysDI levels were increased in dairy rations and where milk protein and casein contents were improved by 1.0 to 2.0 g/kg (Rulquin, Delaby and Hurtaud, 1994). The different protein sources fed in these trials were maize gluten meal, soyabean/rapeseed meal and bloodmeal. Cheese yield was increased in line with the increase in casein content. In terms of cheese-making aptitude there was a tendancy to decrease the time of firming and to enhance firmness. However, curd coagulation time increased significantly, apparently due to a decrease in the colloidal calcium/casein ratio. In practice this can be rectified by increasing the quantity of $CaCl_2$ added to milk before cheese manufacture.

Certain authors have indicated that the different milk casein fractions are improved disproportionately (Donkin, Varga, Sweeney and Muller, 1989) with increasing levels of metDI and lysDI, but this has not been observed in the majority of other trials (Hurtaud et al., 1995; Colin-Schoellen, Laurent, Vignon, Robert and Sloan, 1995). Overall, balancing rations in lysDI and metDI have shown that it is possible to positively manipulate milk protein secretion through increasing milk casein synthesis. This not only provides more grams of casein to be transformed into cheese but also improves the manufacturing properties of the milk.

The first practical approach to formulating in digestible amino acids for dairy cows

The biggest breakthrough recently in practical amino acid nutrition of dairy cows was the compilation of all the relevant research work and the development of a systematic approach to estimating raw material values for digestible methionine and lysine contents and the publication of estimates for digestible lysine and methionine requirements. The methodology used has been described by Rulquin and Vérité (1993), Rulquin, Pisulewski Verite and Guinard (1993) and the previous chapter (Vérité, 1997).

Using this methodology, lysine and methionine requirements for the lactating dairy cow are estimated to be 7.3 % and 2.5 % of total digestible amino acid supply (PDIE) respectively. Socha and Schwab 1994, using a similar approach but based on the Cornell Net Carbohydrate and Protein system (O' Connor, Sniffen, Fox and Chalupa, 1993), estimated requirements for lysine as 16.8 % of total essential amino acids. Specific dose response studies (Robert *et al.,* 1996) have confirmed the estimation of the lysDI requirements derived by Rulquin and Vérité (1993). A precise estimate of metDI requirements has been much more difficult to establish. The marginal response to additional metDI remained positive and linear between 1.5 to 2.35 (% of PDIE) metDI where the ration provided recommended levels of lysDI (7.3 % of PDIE) (Pisulewski *et al.*, 1996; Socha, Schwab, Putnam, Whitehouse, Kierstead and Garthwaite, 1994a,b). However, Socha *et al.* (1994c) found no response to increased metDI in mid to late lactation cows. Where the lysDI level was only 6.50 % of PDIE no significant increase in milk protein secretion has been noted from additional metDI, whereas increasing simultaneously lysDI and metDI from 6.50/1.80 to a minimum of 6.80/2.15, increased milk protein secretion by 37 g/cow/day (Sloan, Robert and Lavedrine, 1994).

The estimates of 2.5 % and 7.3 % of PDIE proposed by Rulquin and Vérité (1993) would, however, appear to be very good first approximations of metDI and lysDI requirements respectively. In countries where this approach to rationing cows in terms of lysDI and metDI as a % of PDIE has been adopted, a reasonable compromise between optimising milk protein secretion and the cost of doing so, has resulted in the adoption of practical recommendations of 7.0 lysDI and 2.2 metDI (as a % of PDIE).

No major interference of other formulation constraints

A question often posed is how dependent is the response to improving digestible lysine and methionine levels on other ration formulation constraints and dietary ingredients present. The responses need to be independent, or at least predictable, in a wide range of dietary situations if lysDI/metDI formulation constraints are to be incorporated in current dairy ration formulation programmes.

RESPONSES LARGELY INDEPENDENT OF RATION ENERGY LEVEL AND SOURCE

In pigs, lysine requirements are expressed with respect to net energy (Fuller,1994); energy being needed for the process of tissue protein synthesis. Erfle and Fisher (1977) commented that an adequate energy intake also had to be provided in dairy cows to permit a response to intravenous infusions of lysine and methionine. However, where energy intakes have been varied between 90 and 110 % of conventional recommendations there has been no noticeable influence on the magnitude of the response to additional digestible methionine and lysine (Rulquin and Delaby, 1994b; Brunschwig, Augeard, Sloan and Tanan, 1995). In fact, Colin-Schoellen *et al.* (1995) observed that the improvement in milk protein secretion to extra metDI and lysDI tended to be higher (88 vs 41 g/cow/day) at low versus high energy intakes (90 vs 112 % of recommendations).

In trials where the extra energy was supplied in the form of fat (Christensen, Cameron, Clark, Drackley, Lynch and Barbano, 1994 ; Canale, Muller, McCahon, Whitsel, Varga and Lormore, 1990) the effects on milk protein content of supplying extra metDI and lysDI were significant and comparable to those obtained on a diet with no added fat. The general negative effect of added fat, independent of source, on milk protein content was counterbalanced by the positive effect of providing extra digestible lysine and methionine.

In one trial with Jersey cows, milk protein secretion was not modified when additional metDI and lysDI were supplied, independently of the presence or not of supplementary dietary fat (Karunanandaa, Goodling, Varga, Muller, McNeill, Cassidy and Lykos, 1994). Nevertheless, in very early lactation Jersey cows (Bertrand, Pardue and Jenkins, 1996) the addition of whole cottonseed depressed milk protein content by 3 g/kg. Adding supplementary metDI and lysDI inversed this decrease completely (4.6 g/kg), improving milk protein yield by 120 g/cow/day. Chow, DePeters and Baldwin (1990) confirmed that with high energy diets, whether attained by the inclusion of fat or by the feeding a high proportion of concentrate, milk protein yields were increased similarly when additional metDI and lysDI were supplied.

OVER AND ABOVE MINIMUM RECOMMENDATIONS FOR TOTAL DIGESTIBLE AMINO ACID SUPPLY (PDI) RESPONSES ARE LARGELY INDEPENDENT OF RATION PROTEIN LEVEL

The influence of the level of the protein supplied (PDI) has also been investigated. Rulquin *et al.* (1990) were able to demonstrate that dairy cows fed to meet 120 % of their recommended PDI requirements were still able to increase milk protein secretion by 150 g/cow/day with the infusion of one single amino acid, lysine. This ration was particularly imbalanced in lysine with respect to other amino acids, due to the high inclusion of maize gluten meal. When, in the same experiment, cows were fed only 100 % of their recommended PDI requirements the response to extra lysine was only 45 g/cow/day. The inclusion level of corn gluten meal was much lower, thus the relative deficit in lysDI was also lower. This was a clear indication that at the levels of protein input examined in this trial it was the relative imbalance in the profile of amino acids being absorbed by the dairy cow that was important in determining the potential response rather than the quantity of total amino acids absorbed.

Where different protein levels have been tested (100 vs 105 % of the recommendation for PDI) and the amino acid profile of protein entering the small intestine at each level was estimated to be similar, milk protein increases to a provision of extra lysDI and metDI were of the same magnitude (Sloan, Robert and Mathé, 1989; Socha *et al.* 1994; Colin-Schoellen. *et al* 1995).

FORAGE SOURCE HAS LITTLE INFLUENCE ON AMINO ACID LIMITATIONS

A major concern in Northern Europe has been that in the original database used to develop the estimates of lysDI and metDI recommendations, very few rations were based on grass silage. Would the recommendations proposed by Rulquin and Vérité (1993) be applicable for this type of diet? Earlier work where lysDI and/or metDI levels have been increased in grass-silage based diets have given very variable results (Girdler, Thomas and Chamberlain, 1988a,b; Thomas, Crocker, Fisher, Walker and Reeve, 1989; Remond, 1988). The increases in milk protein secretion were not only more variable but apparently inferior to those obtained with maize-silage based diets (Le Henaff, 1991). However, other constraints such as low dietary protein levels may have limited the capacity to respond in these trials (Girdler, 1988 a,b; Chamberlain and Thomas, 1982).

More recently, 3 trials have been carried out specifically to test the validity of the lysDI and metDI recommendations, where grass silage was the sole forage fed. In two trials, the control rations were formulated to contain approximately

7.0 lysDI and around 1.80 metDI (% of PDIE). In both trials (Robert, Sloan and Denis, 1994; Chilliard, Rouel, Ollier, Bony, Tanan and Sloan, 1995) milk protein content was improved by 1.2 g/kg which is similar to the responses (1 to 1.5 g/kg) observed with maize silage and mixed forages using similar protocols (Sloan, 1993; Brunschwig *et al*, 1995).

The control ration used in the third trial was based on grass silage and byproducts and was estimated to provide 6.70 lysDI and 1.85 metDI as a % of PDIE (Younge, Murphy, Rath and Sloan, 1995).

To test the approach of Rulquin and Vérité (1993):

1) Only additional digestible methionine was included to achieve a ration level of at least 2.2 % metDI

2) Additional digestible methionine and lysine were added to achieve levels of at least 7.0 % lysDI and 2.2 % metDI

3) The ration was reformulated to achieve the 7.0 % lysDI uniquely from raw material sources, the deficit in metDI being supplied by a post-ruminal source of synthetic methionine.

As was anticipated, adding methionine alone showed virtually no increase in milk protein. Simultaneously adding digestible lysine with the additional digestible methionine permitted a significant increase in milk protein content (+ 0.8 g/kg). The largest increase was achieved by the reformulated diet (+ 1.4 g/kg).

Using the approach of Rulquin and Vérité (1993), Armentano, Bertics and Ducharme (1993) also observed similar improvements in milk protein with a lucerne-based ration, and Robert and Lavedrine (1995) with a grass-hay based ration when the predicted deficit in metDI was rectified. This tends to confirm the robustness of the sytem and the recommendations proposed by Rulquin and Vérité, (1993).

CAN OTHER AMINO ACIDS BE LIMITING ON A PRACTICAL BASIS IN DAIRY COW RATIONS

Obviously balancing dairy rations uniquely for lysine and methionine is not sufficient to achieve maximum milk performance. The comparisons of infusions of essential amino acids versus casein showed that the first two limiting amino acids could account for 40 to 50 % of the total response to casein. Thereafter other amino acids become limiting. Various amino acids have been suggested as being potentially 3rd or 4th limiting. Histidine appears to be a good candidate. Its theoretically low concentration in microbial protein would apparently suggest that mammary supplies of histidine for milk protein synthesis often look limited.

In the early work of Schwab (1976) although lysine and methionine were clearly first limiting, infusion of extra histidine, in addition, tended to increase milk protein secretion through a milk volume response. Recently some Finnish work (Huhtanen, Vanhaatalo and Varvikko, 1996) indicated the potential importance of histidine as a first limiting amino acid on a grass silage, cereal/sugarbeet pulp based diet where urea was the only supplementary source of crude protein.

Another amino acid often thought to be potentially limiting on grass silage/ hay based diets is leucine. In North European rations, compared with American rations which contain a large proportion of maize grain, the proportion of leucine in duodenal protein from cows fed grass-silage based rations has been shown to be up to 6 % lower (Robert, Tanan, Blanchart, Williams and Phillipeau, 1994). However although this may indicate that leucine is potentially 3rd or 4th limiting, methionine and lysine are still clearly first limiting in grass-silage based rations as indicated by the positive milk performance responses observed by Robert *et al* (1994), Chilliard *et al.* (1995), and Younge *et al.* (1995) to increasing digestible supplies of just these two amino acids.

Isoleucine has also been suggested as a potential limiting amino acid. The Cornell Net Carbohydrate and Protein system often predicts isoleucine as the first limiting amino acid. However this is probably an artefact of the low transfer coefficient (0.79) of absorbed leucine to leucine incorporated into milk protein, used in the model, compared with 0.98 and 0.88 for methionine and lysine respectively. There are no milk performance trials indicating isoleucine as a potentially limiting amino acid.

The non - essential amino acids are normally discounted as being potentially limiting for milk protein synthesis. However, Bruckental, Ascarelli, Yosif and Alumot (1991) indicated that proline could potentially increase milk yield by having a sparing effect on arginine which may contribute up to 50 % of the proline secreted in milk. Meijer, Van der Meulen and Van Vuuren (1993) proposed that glutamine could be potentially limiting, particularly in early lactation. The plasma concentration of this amino acid is extremely low compared to other amino acids at this critical period of lactation. However as yet there is no direct evidence that increasing digestible supplies of glutamine can increase any aspect of milk performance.

In practice, as we do not yet formulate to fully meet the recommendations for the two first limiting amino acids lysine and methionine, there is little point in trying to incorporate formulation constraints for the 3rd and 4th limiting amino acids based on the limited knowledge we have available today. However it is important for the future that the sequence of limitations is more clearly identified. In pigs the identification of threonine and tryptophan as 3rd and 4th limiting amino acids has given the pig nutritionist the opportunity to reduce crude protein (CP) inputs whilst maintaining performance.

The largest benefits in milk performance are achieved from balancing rations for lysDI and metDI from the beginning of lactation

BENEFITS OF CONTINUOUS APPLICATION FROM EARLY IN LACTATION

The trials used by Rulquin and Vérité (1993) to develop their recommendations virtually all used a latin square or crossover design carried out in post-peak lactation with periods sometimes as short as two weeks. Broster and Broster (1984) pointed out the dangers of directly extrapolating results of this nature into recommendations for a whole lactation. The responses to long term improvements in lysDI and metDI contents of dairy rations could as easily increase as disappear with time. The benefits could also be quite different depending on the stage of lactation at which the ration is first optimised for lysDI and metDI content. For example, Rulquin (1992) noted that in early-lactation trials, milk protein secretion responses to supplementary lysine and methionine were apparently greater than in later lactation studies (50 vs 35 g/cow/day). Schwab, Bozak, Whitehouse and Olson (1992) also noted in a repeated Latin square study, where the composition of the total mixed ration was virtually kept constant, that the response to additional lysDI and metDI was greater during the first 50 days compared with any other stage of lactation. Increases in both milk protein content (~ 2.6 vs ~ 1.3 g/kg) and milk yield (2.4 vs 0.0 kg/day) were much larger in very early lactation than any period post-peak lactation.

Brunschwig and Augeard (1994) observed that where a ration estimated to have a concentration of 6.9 lysDI and 1.75 metDI was readjusted to have 2.15 metDI (% of PDIE) from the 4th week of lactation, milk protein secretion was improved by 60 g/cow/day within a matter of days and this response was maintained at this level over the length of the trial period (until the 20th week of lactation - Figure 10.1a). This response is larger than those observed in the cross over trials summarized by Rulquin (1992). The increase in milk protein concentration (Figure 10.1b) was at first sight unexpected. An initial increase of 1 g/kg was observed at the first milk recording but by the end of the trial the difference had steadily increased to nearly 2 g/kg. This was due principally to the increased protein secretion becoming more and more concentrated in a decreasing volume of milk, post peak lactation. In addition there was an indication that milk yield was improved slightly (⊢ 0.5 kg/cow/day) during the first few weeks of the trial.

In a further trial (Pabst unpublished) where ration lysDI and metDI concentrations as a % of PDIE were increased from 6.8/1.9 to 7.4/2.1, milk protein secretion was increased immediately by 75 g. This level of response was maintained until the end of the trial at the 30th week of lactation. Here the positive effect on

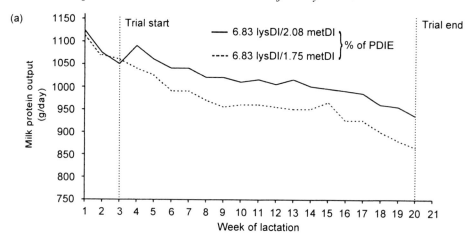

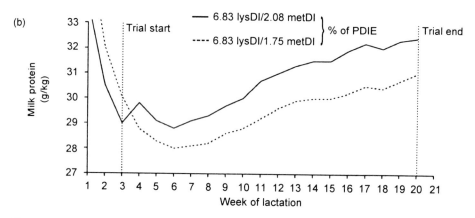

Figure 10.1 Change in milk protein output (a) and content (b) from week 4 to 20 of lactation after optimizing ration for lysDI and metDI (% of PDIE) (Brunschwig and Augeard, 1994)

milk protein content over time was even more pronounced than in the previous experiment; 1 g/kg at the beginning of the trial, 2.8 g/kg at the end. Polan, Cummins, Sniffen, Muscato, Vicini, Crooker, Clark, Johnson, Otterby, Guillaume, Muller, Varga, Murray, and Peirce Sandner (1991) noted an average increase of 70 g per day in milk protein secretion during days 22 to 280 of lactation when cows fed maize gluten meal as the principal dietary protein source were supplemented with extra metDI and lysDI. It appeared that the average daily increase was larger (+ 85 g) during days 22 to 112 of lactation than subsequently (+ 40 g). However, with a similar trial design and ration, Rogers, Peirce-Sandner, Papas, Polan, Sniffen, Muscato, Staples and Clark (1989) observed an increase of over 90 g/day in milk protein secretion, which was maintained throughout the whole lactation. Thus there does seem to be an advantage to optimising lysDI/metDI levels from as early a stage of lactation as possible, not only to obtain a larger immediate effect, but also to stimulate the largest response possible over the whole lactation.

MILK VOLUME INCREASES IN EARLY LACTATION

The trials discussed previously did not start before the 4th week of lactation to provide a covariate period. It could be argued that the most stressful time of lactation has already passed. The dairy cow may, in fact, benefit even more if her amino acid nutrition is improved from Day 1 of lactation or even pre-partum.

In a recent trial involving 72 high yielding dairy cows (Socha, Schwab, Putnam, Whitehouse, Kierstead and Garthwaite, 1994d), a typical North American diet, which contained 6.50 lysDI / 1.85 meDI, was compared with a ration in which the lysDI/metDI levels were increased to 7.10 / 2.15 (as a % of PDIE) respectively from parturition until Day 105 of lactation. Milk protein yield was increased by an average of 80 g/cow/day, which, as in the trials of Brunschwig and Augeard 1994 and Pabst unpublished, was maintained consistently throughout the trial period. In the trial of Socha, *et al* (1994) the expression of the milk protein yield response was principally in terms of a milk volume response (+ 2.3 kg/day on average, + 3.5 kg/day at peak lactation) although there was some evidence by the end of the trial that the milk volume response was becoming less important and the protein content increase more important (Figures 10.2a and b).

Other trials (Rulquin, personal communication; Chilliard Ollier, Ferlay, Gruffat, Durand and Bauchart unpublished) have confirmed this large potential milk volume response where rations have been balanced for lysDI/metDI from the first week of lactation. The latter trial was also associated with a large increase in milk protein content. A further trial (partly published ; Robert, Sloan and Bourdeau, 1994) also demonstrated a large effect on milk protein yield during the first 84 days of lactation, the majority of the response being due to an increase in milk protein content (1.6 g/kg) with only a small effect on milk yield (+ 0.5 kg/day). A recently published farm study (Thiaucourt, 1996) involving 2000 cows also confirmed that balancing rations to try and achieve an objective of 7.0 lysDI / 2.2 metDI does indeed improve milk yield during the first 100 days of lactation (+ 2.2 kg/day) with the increase in milk protein content being evident (+ 1.3 g/kg) at any stage of application during the lactation.

In cows fed rations balanced for lysDI and metDI from at least the first day of lactation, milk protein secretion is increased by 60 to 100 g/cow/day. The expression of this response, which appears to remain relatively constant throughout a large part of lactation, is associated partly with a milk yield response during at least the first 100 days and an increase in milk protein content at most stages of lactation.

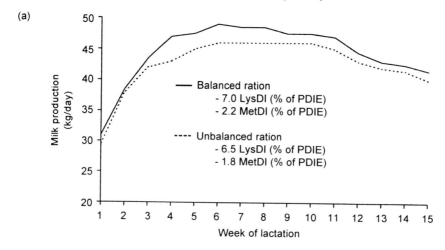

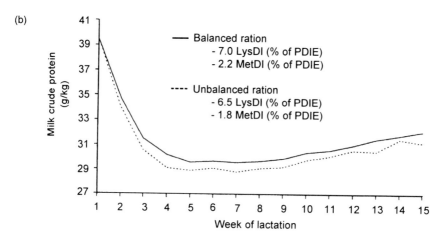

Figure 10.2 Influence of balancing rations to 7.0 lysDI and 2.2 metDI (% of PDIE) on milk yield (a) and milk protein content (b) during the first 100 days of lactation (Socha *et al.*, 1994d)

A POSITIVE INFLUENCE ON EFFICIENCY OF FEED NITROGEN AND ENERGY UTILISATION

In the trials of cross over design started post peak lactation, summarized by Rulquin (1992) the responses in daily milk protein output to supplying extra metDI and lysDI were 20 to 45 g. On average, this was achieved without a modification of feed intake. Thus the global efficiency of conversion of feed crude protein to milk protein can be estimated to be improved by the order of 0.02 to 0.05, which is not neglible. However, when the proportion of the extra metDI appearing in milk protein is calculated, it averages only 0.1. Theoretically the efficiency of utilisation for the first limiting amino acid could approach 1. However this is not realistic.

Methionine is used not only for milk protein synthesis but also for maintenance functions and is implicated in other metabolic processes. Furthermore, after peak lactation a proportion of the extra metDI may be directed towards tissue protein synthesis. Realistically it cannot be expected that more than 0.6 to 0.8 of a supplementary quantity of the first limiting amino acid will be incorporated into milk protein.

Similarly, in the early lactation studies (Robert, *et al* unpublished; Brunschwig and Augeard, 1994; Socha *et al.*, 1994d), improving amino acid balance had no overall effect on feed intake. However, for the same quantities of additional metDI and/or lysDI the increases in milk protein secretion were much greater (60 to 80 g/cow/day), equivalent to an improvement in effiency of feed protein use of 0.06 to 0.10. Equally the efficiency of utilisation of the supplementary quantity of the first limiting amino acid, methionine, was higher. For Brunschwig *et al.* (1995), the extra metDI was used with an efficiency of 0.2. In the trials of Pabst (unpublished) and Socha *et al.* (1994d), approximately 0.4 of the extra metDI was accounted for in extra milk protein secreted.

What is more curious is the apparent influence of balancing rations for lysDI and metDI on overall feed/energy utilisation. For 12 early-lactation trials analysed there was an improvement in apparent feed efficiency ((feed DM input/milk DM output (milk fat + milk protein)) in every case (Table 10.2). This could not be explained solely by an increase in milk protein output. In the trials where milk volume was improved, milk fat content was not reduced, thus milk fat secretion was also increased. In fact milk component yield (fat plus protein) was often increased by at least 150 g/cow/day with an increase in apparent feed efficiency of 0.04 to 0.08. Ørskov, Grubb and Kay (1975) showed that feeding fishmeal, a protein capable of providing bypass protein of a high quality, caused an increase in milk yield and milk protein secretion at the expense of mobilising more energy reserves to satisfy the increased energetic demand.

In the early lactation trials cited above there was no indication of more weight or condition score loss (where measured). This would tend to favour a hypothesis that energy yielding nutrients both of dietary origin and from body reserves are being utilised much more efficiently for milk production when rations are balanced for lysDI and metDI.

POSITIVE EFFECTS ON REPRODUCTIVE PERFORMANCE

It can be anticipated that reproductive performance of dairy cows may be influenced in various ways by balancing more correctly lysDI and metDI supplies. Any feeding practice that allows cows to return to positive energy balance sooner and/or reduce blood urea levels (Ferguson and Chalupa,1989) will create a more favourable

Table 10.2 INFLUENCE OF RATION LYSDI/METDI CONCENTRATION ON MILK PERFORMANCE AND FEED EFFICIENCY

Reference	Trial duration (days of lactation)	Ration lysDI/metDI concentration (% PDIE)		Increase in intake (kg DM/cow/day)	Increase in milk fat plus protein yield (g/cow/day)	Apparent feed efficiency kg DM intake/kg milk DM (fat plus protein output)	
		Control	Supplemented	(Supplemented-Control)		Control	Supplemented
Robert et al, unpublished[†]	0-84	8.0/1.7	6.9/2.4	-0.1	131	7.35	6.92
Brunschwig et al 1994	21-140	6.9/1.8	6.9/2.4	0.4	149	8.67	8.29
Brunschwig et al 1995[†]	0-56	6.9/1.7	6.9/2.1	-1.1	47	7.27	6.72
Pabst unpublished[†]	21-203	6.8/1.9	7.4/2.1	0.7	187	7.86	7.50
Socha et al 1994d[†]	0-105	6.5/1.8	6.5/2.2	-1.0	-32	9.06	8.80
	0-105	6.5/1.8	7.0/2.2	0.8	158	9.06	8.85
Rulquin unpublished[†]	0-28	6.9/1.7	7.7/2.1	0.4	146	7.48	7.21
Chilliard et al unpublished[†]	5-42	6.9/1.6	6.9/2.2	0.1	291	6.48	5.87
	5-26	6.9/1.6	6.8/2.8	1.1	269	6.20	5.99
	5-26	6.9/1.6	8.2/2.6	1.7	133	6.20	6.56
Robinson et al 1995	27-305	6.2/1.9	6.8/2.2	1.0	150	10.73	10.40
	27-305	6.2/2.0	6.8/2.2	0.0	120	10.84	10.27

[†] Actual milk production values - no adjustment by covariance

environment for getting cows back in calf. Balancing rations in lysDI/metDI has the potential to achieve these objectives. Due to the key role that methionine plays in energy metabolism (liver function) in early lactation, cows fed diets fortified with metDI should be in a much more favourable energy status for successful reproduction. In two recent studies Robert *et al.* (1996) found tendancies to reduce calving interval by 4 to 5 days on both maize- and grass-silage based rations where metDI levels were increased from 1.8 to 2.2 % of PDIE. In parallel field trials, Thiaucourt (1996) found the same improvement in calving interval ($P < 0.1$) in a field study involving 2000 cows that were fed a metDI-enriched concentrate.

These advantages are a logical reflection of improving energy and protein nutrition in general without necessary implicating a favourable effect of one individual amino acid on any direct aspect of reproductive function.

In the studies of Robert *et al.* (1996), the reduction in calving interval was due essentially to two factors:

1)   Number of days to first insemination was reduced by 6 to 7 days
2)   The number of inseminations needed per successful conception was reduced from 1.8 to 1.6

Both could be linked to the time taken for full uterine involution after calving. The proportion of cows fed the metDI-enriched diet that had completely involuted after 45 days of lactation was 0.55, compared to only 0.38 of the control animals.

Milk progesterone levels were also measured every three days during this study. Cows fed the metDI enriched diets had higher progesterone levels (16.5 vs 12.6 nmol/litre) during the 5 to 7 days prior to the ovulation leading to a successful pregnancy. This is generally considered to be favourable to a strong ovulation and successful insemination (Folman, Rosenberg, Herz and Davidson, 1973). Subsequently after insemination higher progesterone levels are also favourable for implantation of the embryo. Robert *et al.* (1996) also observed milk progesterone levels were on average higher (4.8 vs 7.1 nmol/litre ; $P < 0.1$) during the five days post ovulation for the metDI enriched diets.

PRE-CALVING AMINO ACID NUTRITION IMPROVES SUBSEQUENT LACTATION PERFORMANCE ?

During the weeks prior to calving, the dairy cow undergoes important hormonal and metabolic changes in preparation for calving and in anticipation of secreting large quantities of milk. Nutrition during this time is vital to minimising subsequent metabolic disorders and ensuring a rapid increase in intake post-calving to help satisfy the large demand for nutrients in early lactation (Grummer, 1996). Labile protein reserves can also be replenished during this time (Chilliard and Robelin, 1983) to provide a small reservoir to help meet amino acid requirements at the beginning of lactation.

As with lactation rations, it would appear that in the pre-calving ration there is first and foremost a need to ensure provision of a minimum quantity of total amino acids (PDI) which has perhaps been underestimated in the past. Moorby, Dewhurst and Marsden (1996) have recently shown that increasing total digestible amino acid supply by approximately 100g/day during the 2 months before calving increased milk protein secretion post calving by 60 g/cow/day. There was no dietary difference post calving and intakes were similar. From where did the extra milk amino acid N secreted originate? One hypothesis is that pre-calving, the extra amino acid N fed was used in some way to constitute labile protein reserves. However, it has traditionnally been considered that the potential contribution of labile protein reserves to meet amino acid requirements is exhausted after the third or fourth week of lactation (Chilliard and Robelin, 1983). It would therefore seem likely that by some mechanism, the increased total amino acid supply pre-calving improves subsequently the efficiency of utilisation of absorbed amino acids for milk protein synthesis.

According to Rode, Fayceda, Sab, Suzuki, Julien and Sniffen (1994) attention should also be paid to the balance between rumen degradable N and bypass protein supplied in the dry cow ration. They found that feeding too high a proportion of bypass protein could in itself have potential negative effects on subsequent health and production. Both Crawley and Kilmer (1996) and Wu, Fisher and Polan (1996) showed no advantage of increasing either the proportion of crude protein in the dry cow ration or the proportion of undegradable protein in the crude protein on subsequent lactational performance.

It could be anticipated that pre-calving there would be an advantage in terms of the efficiency of utilisation of absorbed amino acid N, if the profile approached that of 'ideal protein'. Both Socha *et al.* (1994) and Rode *et al.* (1994) observed very large responses in milk protein yield, during the first hundred days of lactation, to feeding additional metDI and lysDI. Part of this response may be attributed to the additional quantities of metDI and lysDI which were also fed pre-calving.

Further evidence that not only the quantity but the balance of amino acids absorbed by the dairy cow during the dry period is important in determining lactational performance has been demonstrated by Socha *et al.* (1994), Robert *et al.* (unpublished) and Brunschwig *et al.* (1995). In these 3 trials the quantity of metDI calculated to meet the estimated deficit in the lactation ration was also fed for at least the last two weeks before parturition as part of the trial protocol. The results were not as anticipated since there was no increase in milk yield in any of the 3 trials In fact there was a decrease of 1.2 kg/day in the trial of Brunschwig *et al.* (1994), although the yield of milk components did not change. The trial of

Robert *et al.* (unpublished) was carried out over two years. Overall milk component yield was increased by 140 g/cow/day. However, the results were different between the years. In year 1, a positive effect on milk volume was noted (+ 1.1kg/cow/day) whereas the effect was negative in the second year. For the trial of Socha *et al.* (1994) the addition of supplementary metDI alone was not expected to increase markedly milk performance. Although methionine was estimated to be the first limiting amino acid for the ration fed, lysDI (6.50 % of PDIE) was well below an optimum level to ensure a full response to the additional metDI. However it was surprising that there was no response at all.

The common feature of all these trials was a negative effect on feed intake (kg DM/head/day) during early lactation (Brunschwig *et al.*, 1995,- 1.1; Robert *et al.*, unpublished, - 0.5 in the second year; Socha *et al*, -.1.0 during the first 100 days of lactation). These negative effects were evident from the first day after parturition. In fact, the problem was apparently initiated pre-calving. In all 3 trials, metDI levels pre-calving were higher than those found in lactation rations (2.4 to 3.1 as a % of PDIE) due to the feeding of a supplementary level of metDI designed for cows in lactation consuming higher quantities of PDIE. In the two trials where feed intake was measured before calving, there was no difference between treatments until 24 to 48 hours before calving. Intakes normally decrease around this time but in the case of Socha *et al.* (1994d) the decrease was spectacular for the treatment with added metDI - (Figure 10.3) and also important in the trial of Robert *et al.* (unpublished).

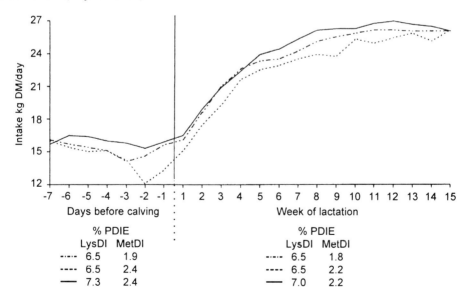

Figure 10.3 Influence of ration lysDI and metDI concentrations (% of PDIE) on dry matter intake before and after calving (Socha *et al*, 1994d; Schwab personal communication).

It would appear that it is not only the level of metDI as a proportion of total amino acids (PDIE) that is important, but also the ratio of lysDI to metDI. In the trial of Socha *et al.* (1994) the feeding of a high level of metDI in the presence of a high level of lysDI had no impact on intake in the period one to two days before calving and appeared to positively favour milk performance in the subsequent lactation. Grummer (1996) showed quite clearly that intakes post-calving, and thus lactational performance, were positively correlated with intakes immediately pre-calving. Thus it is evident that there is no interest in exceeding the normal recommendations for metDI and lysDI as a proportion of PDIE pre-calving. Despite the negative effects incurred on intake, the apparent efficiencies of utilisation of feed nitrogen and energy were still improved.

It is important to achieve a balanced profile of digestible amino acids when formulating a dry cow ration. The quantities of individual digestible amino acids necessary to balance a ration should be calculated with respect to the overall digestible protein supply, which is only approximately 0.5 of protein supply during the early lactation period.

Methionine, a key role in liver energy metabolism

The positive effects on feed utilisation cannot be explained totally by improvements in milk protein synthesis. In contrast to lysine, which is considered to have a role in protein synthesis only, methionine is known to participate in many other metabolic functions through its capacity as a methyl donor. Recently its role in liver energy metabolism has been investigated. Chilliard *et al.* (unpublished), over fed dairy cows precalving to be fat at calving and then fed an energy-deficient diet post-calving. These cows were predisposed to develop metabolic problems such as fatty liver (steatosis) and ketosis. The control diet was formulated to provide ample PDI and to meet practical requirements for lysDI (7.0 % of PDIE), but was deficient in metDI. For the test rations, additional metDI was provided to achieve a level of 2.2 % of PDIE in trial 1 and 2.8 % of PDIE in trial 2. Substantial increases in protein secretion (+ 100 g) were observed in both trials. In trial 2 this was associated with a significant reduction in plasma ketone body concentrations. Plasma ß-hydroxybutyrate and acetone concentrations were reduced by 0.33 and 0.64 respectively during week 2 of lactation, indicating fewer potential problems of energy metabolism and improved liver function. These improvements were assumed to be the result of the increased supply of metDI mediating an increase in very low density lipoproteins (VLDL), the lipoprotein complex essential for evacuating mobilised tryglycerides from the liver towards peripheral tissues, particularly the mammary gland, to provide energy nutrients for lactation.

Auboiron, Durand, Robert, Chapman and Bauchart (1995) demonstrated in calves that methionine plays a key role in assuring the synthesis of the apo protein ß complex, essential in the formation of VLDL. Likewise Durand, Chilliard and Bauchart (1992) showed in lactating cows that increasing liver supply of methionine in the presence of additional lysine resulted in the VLDL balance across the liver becoming positive rather than negative. Chilliard, Audigier, Durand, Auboiron and Zauchart (1994) observed that the supplementary metDI also decreased circulating levels of not only ß- hydroxybutyrate but also plasma triglycerides and non esterified fatty acids (NEFA) in calves. In dairy cows NEFA concentrations were also reduced in the trials of Pisulewski, Rulquin and Vérité (1996) where a ration adequate in lysDI (6.9 %) but poor in metDI was supplemented with additional amounts of metDI. Similarly, in the trial of Rulquin and Delaby (1994a), plasma NEFA were decreased from 96.2 to 74.7 and from 60.9 to 47.0 mmol/l for animals in low and normal energy status respectively, when additional metDI was added. Plasma NEFA are generally positively correlated with the level of lipid infiltration of the liver (Durand, Gruffat, Chilliard and Bauchart, 1995). Fortifying diets with metDI appears to be a positive factor towards reducing the risk of certain metabolic disorders such as steatosis and ketosis and improving the energy status of dairy cows in early lactation.

Potential to reduce nitrogen pollution

An improvement in feed nitrogen utilisation opens up the possibility of reducing urinary nitrogen excretion. This is of importance in countries where legislation is being introduced to control nitrogen usage on farms to reduce the potential negative effects of excess nitrogen on the environment. Losses through urinary excretion have been shown to increase rapidly (Wright, Moscardini, Susnel and McBride, 1996) with increasing dietary crude protein concentration, even when high bypass protein sources, with an apparently good amino acid profile, were fed. It has been shown in pigs that crude protein concentrations in rations can be reduced from 160 to 130 g/kg, without loss in performance, by judicious supplementation with synthetic amino acids. In dairy cows the same potential should exist.

At present rations for dairy cows tend to have a crude protein concentration of over 180 g/kg in the dry matter and sometimes higher. This is because it is generally recognised that there continues to be a marginal increase in milk production with increasing crude protein level (Gordon, 1980). Using the PDI, lysDI / metDI approach, reducing dietary crude protein levels to 160 g/kg DM can reduce crude protein inputs (Dinn, Fisher and Shelford, 1996) by more than 500 g/cow/day. Any marginal loss in milk yield is more than offset by the positive effect of amino acid balance on milk protein secretion (Robert, Sloan, Saby, Mathé, Dumont and

Dzyzcko, 1989). However, to go further and reduce crude protein level by another 20 to 30 g/kg will be much more difficult, though not infeasible. Piva Masoero, Mancini and Fusconi (1996) fed a low crude protein diet (153 g/kg DM) to dairy cows from days 30 to 90 of lactation. With this diet, milk protein secretion was improved (+ 50 g/cow/day) when additional metDI and lysDI were included. However the increase was only 0.5 of that achieved by increasing crude protein concentration to 173 g/kg DM although the latter was at the expense of increasing milk urea N by 0.25 and reducing the efficiency of feed N utilisation. Likewise, Piepenbrink, Overton, and Clark (1996) found that supplementing a ration containing 140 g CP/ kg DM with only digestible lysine and methionine did not have any effect on milk protein secretion, whereas increasing the crude protein level to 180 g/kg DM increased it by 40 g/cow/day.

It seems likely that at low dietary crude protein concentrations more attention needs to be paid not only to lysine and methionine but to other potentially limiting amino acids. In pigs, dietary CP concentrations can be reduced from 160 to 130 g/ kg DM without loss in performance, but only if, in addition to digestible lysine and methionine, the concentrations of the 3rd and 4th limiting amino acids, tryptophan and threonine, are also optimised in the ration.

Practicalities of formulating in lysDI and metDI

All raw materials can contribute to the lysDI and metDI needs of the dairy cow. It is the judicious choice of raw materials and the proportions incorporated into concentrate formulations which will be important in achieving the required overall levels of digestible lysine and methionine.

In the approach proposed by Rulquin and Vérité (1993), each raw material is assigned a lysDI and metDI value which takes into account both the microbial contribution calculated from the quantity of fermentable organic matter (FOM) digested in the rumen and the bypass contribution. All forages have relatively high lysDI values (6.8 to 7.0 % of PDIE) and above average metDI values (1.9 to 2.0 % of PDIE). This is due to the majority of the contribution to lysDI and metDI being of microbial origin. In general, rumen degradabilities of forages are high, thus the digestible amino acid contribution from bypass protein is minor. Likewise for cereals, lysDI (% of PDIE) levels tend to be slightly above average, due to their low crude protein content and high fermentable organic matter content, which again favours the microbial digestible amino acid contribution. Maize grain is an exception. Although its crude protein content is low, rumen N-degradability is very low (~ 0.40) and at least 0.2 of the FOM escapes rumen fermentation. These factors maximise the bypass digestible amino acid contribution for this raw material. With its low lysine content this results in a low value of lysDI (% of

PDIE) for maize and, in general, for maize-based byproducts. Corn gluten meal has the distinction of being the raw material with the lowest lysDI concentration because of its high CP content and extremely low degradability (0.28). On the other hand, maize gluten feed has a lysDI value of 6.4 (% of PDIE) which is average. This is due to the high N-degradability (~ 0.69) of this raw material.

The raw materials that can most influence the profile of absorbed protein (PDIE) are concentrated sources of bypass protein. All protein-rich raw materials, particularly those with a low N-degradability, have the potential to modify cosiderably the amino acid profile of absorbed protein. Soyabean meal is above average in lysDI (7.0 % of PDIE). However all the other proteinaceous meals of vegetable origin tend to be low in both lysDI and metDI. Only animal protein sources have levels of either of the potential limiting amino acids superior to those required in the overall ration. Bloodmeal (now banned in the European Union) is an excellent source of lysDI, but fishmeal is the only raw material that is an excellent source of both lysDI and metDI as a % of PDIE.

In practical ration formulation it is impossible to meet both the lysDI and metDI constraints of 7.0 and 2.2 (as a % of PDIE) respectively from conventional raw materials alone; cows cannot be fed entirely on fishmeal. Unfortunately, feeding synthetic amino acids is not a potential solution. Neither DL-methionine nor the hydroxy analogue of methionine, the two commercially available forms of synthetic methionine, resist rumen degradation (Loerch and Oke 1989) and thus are not sources of bypass methionine to increase ration metDI levels. The recent availability of good quality sources of rumen protected amino acids however not only permits the lysDI and metDI constraints to be met but also gives the formulator greater choice and flexibility in the other raw materials that can be incorporated to attain these levels.

In order to formulate as accurately as possible it is important to have a reliable database of lysDI and metDI values for all the commonly used raw materials. The database issued by INRA (Rulquin Guinard Vérité and Delaby, 1993) provides the vast majority of raw material values needed. For raw materials not present in the tables, and for those recognised to be inherently variable, new lysDI and metDI values can be derived from measuring rumen N-degradability *in sacco* and from measuring the amino acid profile in the original raw materials.

The lysDI and metDI values of rumen protected amino acids cannot be measured accurately in the same way. *In sacco* degradability measurements will tend to over-estimate rumen protection to varying degrees, depending on the coating technology. Present rumen protection technologies rely on physical protection of a core of amino acids by a fat-based coating. This type of coating is more sensitive to physical pressures than chemical breakdown, thus nylon bags can create a protective environment, minimising potential losses of protection by abrasion in the rumen or by mastication. In addition, for many technologies that show excellent rumen protection, it is not evident that the amino acids can be released post-ruminally for absorption in the small intestine.

To overcome these obstacles, a specific methodology has been developed to estimate amino acid bio-availability for rumen protected amino acids. Based on the principle that what is of most importance is the bio-availability of the amino acid(s) to the host animal and not necessarily precise knowledge of the proportion resisting rumen breakdown and the proportion released and absorbed in the small intestine, a quantitative blood methodology has been developped (Rulquin, personal communication). The kinetics of appearance in the plasma of post-ruminally infused methionine is compared to the feeding of a known quantity of rumen protected product to give a quantitative estimation of bio-availability. The various rumen protection technologies tested have revealed a wide range in terms of their effectiveness (proportion of 'protected' methionine actually absorbed by the dairy cow has varied from 0 to 0.8).

It is important that manufacturers of rumen-protected amino acids provide independent measurements of amino acid bio-availability using a standardised methodology as described above. Only then can the technical and economic interest of different rumen protected amino acid technologies as sources of metDI and lysDI in dairy rations be evaluated.

Formulating for digestible amino acids can improve dairy herd performance

Paying more attention to the balance of amino acids absorbed by the dairy cows has the potential to improve milk protein yield and certain other aspects of milk performance:

1. Milk protein content can be increased immediately in cows at any stage of lactation, by 1 to 2 g/kg
2. Milk yield can be improved by up to 2.5 kg/day in early lactation (first 100 days)
3. Milk protein yield increases of 60 to 100 g/ day over a complete lactation are attainable
4. Feed efficiency can be improved by 0.05.

Energy status in early lactation is improved, reducing the incidence of metabolic disturbances such as fatty liver and ketosis. Due to its specific action on the liver, digestible methionine concentrations in the ration are of particular importance at the very beginning of lactation. Furthermore, in the few trials where reproductive parameters have been studied, one of the consequences of improved nutritional status has been a shortening of the calving interval, linked to improved uterine involution and a reduction in the number of inseminations needed per conception.

To optimise these potential benefits it is essential to formulate dairy rations to contain higher levels of both lysDI and metDI than is currently practiced. The approach and recommendations proposed by Rulquin and Vérité (1993) are sufficiently robust that they can be used satisfactorally in a wide variety of dietary situations. The challenges for the future will be to refine these recommendations still further with respect to stage of lactation and milk yield potential and to determine recommendations for the dietary levels of the other essential amino acids in order to improve the efficiency of utilisation of dietary protein inputs .

References

Agricultural and Food Research Council (1992). Technical Committee on Responses to Nutrients. Report n° 9. Nutritive requirements of ruminant animals protein. *Nutrition Abstracts and Reviews*, Series B **62**, 787– 835.

Armentano, L.E. and Bertics, S.J. and Ducharme, G.A. (1993). Lactation responses to rumen-protected methionine, or methionine with lysine, in diets based on alfalfa haylage. *Journal of Dairy Science* **78,** (suppl. 1) 202.

Auboiron, S., Durand, D., Robert, J.C., Chapman, M.J. and Bauchart, D. (1995). Effects of dietary fat and L-methionine on the hepatic metabolism of very low density lipoproteins in the preruminant calf, Bos spp. *Reproduction Nutrition Development*. **35**, 167–168.

Bertrand, J.A., Pardue, F.F., and Jenkins, T.C. (1996). Effect of protected amino acids on milk production and composition of Jersey cows fed whole cottonseed. *Journal of Dairy Science* **79** (Suppl. 1), 5022.

Broster, W.H. and Broster, V. J. (1984). Reviews of the progress of dairy science and long term effects of nutrition on the performance of the dairy cow. *Journal of Dairy Research* **51**, 149–196.

Bruckental, I., Ascarelli, Y., Yosif, B. and Alumot, E. (1991). Effect of duodenal proline infusion on milk production and composition in dairy cows. *Animal Production* **53**, 299–303.

Brunschwig, P. and Augeard, P. (1994). Acide aminé protégé : effet sur la production et la composition du lait des vaches sur régime ensilage de maïs. *Journées de Recherches sur l'alimentation et la nutrition des Herbivores 16-17 mars - INRA Theix.*

Brunschwig, P., Augeard, P., Sloan, B.K., and Tanan, K., (1995). Feeding of protected methionine from 10 days pre-calving and at the beginning of lactation to dairy cows fed a maize silage based ration. *Rencontres Recherches Ruminants* **2**, 249.

Brunschwig, P., Augeard, P., Sloan, B.K., and Tanan, K. (1995). Supplementation of maize silage or mixed forage (maize and grass silage) based rations with rumen protected methionine for dairy cows. *Annales Zootechnie*, **44**, 380.

Canale, C.J., Muller, L.D., Mc Cahon, H.A., Whitsel, T.J., Varga, G.A. and Lormore, M.J. (1990). Dietary fat and ruminally protected amino acids for high producing dairy cows. *Journal of Dairy Science* **73**, 135–141.

Chamberlain, D.G. and Thomas, P.C. (1982). Effect of intravenous supplements of L-methionine on milk yield and composition in cows given silage cereal diets. *Journal of Dairy Research*, **49**, 25–28.

Chilliard, Y. and Robelin, J. (1983). Mobilisation of body proteins by early lactating dairy cows measured by slaughter and D_2O techniques. Proceedings of the fourth international symposium on protein metabolism and nutrition, Clermont-Ferrand (France). *INRA publication II (Les colloques de l'INRA, n° 16)* p 195–198.

Chilliard, Y., Audigier, C., Durand, D., Auboiron, S. and Bauchart, D. (1994). Effects of portal infusions of methionine on plasma concentrations and estimated hepatic balances of metabolites in underfed preruminant calves. *Annales Zootechnie*, **43**, 299.

Chilliard, Y., Rouel, J., Ollier, A., Bony, J., Tanan, K. and Sloan, B.K. (1995). Limitations in digestible methionine in the intestine (metDI) for milk protein secretion in dairy cows fed a ration based on grass silage. *Animal Science,* **60**, 553.

Choung, J.J. and Chamberlain D.G. (1993). The effect of abomasal infusions of casein or soya-bean-protein isolate on the milk production of dairy cows in mid lactation. *British Journal of Nutrition* **69**, 103–115.

Chow, J.M., DePeters, E. J. and Baldwin, R. L. (1990). Effect of rumen-protected methionine and lysine on casein in milk when diets high in fat or concentrate are fed. *Journal of Dairy Science* **73**, 1051–1061.

Christensen, R.A., Cameron, M.R., Clark, J.H., Drackley, J.K., Lynch, J.M. and Barbano, D.M. (1994). Effects of amount of protein and ruminally protected amino acids in the diet of dairy cows fed supplemental fat. *Journal of Dairy Science* **77**, 1618–1629.

Clark, J.H. (1975). Lactational responses to postruminal administration of proteins and amino acids. *Journal of Dairy Science* **58**, 1178–1197.

Colin - Schoellen, O., Laurent, F., Vignon, B., Robert J.C. and Sloan B.K. (1995). Interactions of ruminally protected methionine and lysine with protein source or energy level in the diet of cows. *Journal of Dairy Science* **78**, 2807–2818.

Dinn, N.E., Fisher, L.J. and Shelford, J.A. (1996). Using rumen-protected lysine and methionine to improve N efficiency in lactating dairy cows by reducing the percentage of intake N excreted in urine and feces. *Journal of Animal Science* **74**, (Suppl. 1) 11.

Donkin, S.S., Varga, G.A., Sweeney, T.F., and Muller, L.D. (1989). Rumen-protected methionine and lysine: effects on animal performance, milk protein yield, and physiological measures. *Journal of Dairy Science* **72**, 1484–1491.

Durand, D., Chilliard, Y. and Bauchart, D. (1992). Effects of lysine and methionine on in vivo hepatic secretion of VLDL in the high yielding dairy cows in early lactation. *Journal of Dairy Science*, **75**, (Suppl. 1) 279.

Durand, D., Gruffat, D., Chilliard, Y. and Bauchart, D. (1995). Stéatose hépatique: mécanismes et traitements nutritionnels chez la vache laitière. *Le Point Vétérinaire* **27**, 741–748.

Erfle, J.D. and Fisher, L.J., (1977). The effects of intravenous infusion of lysine, lysine plus methionine or carnitine on plasma amino acids and milk production of dairy cows. *Canadian Journal of Animal Science* **57**, 101–109.

Ferguson, J.D. and Chalupa, W. (1989). Symposium : Interaction of Nutrition and Reproduction. Impact of protein nutrition on reproduction in dairy cows. *Journal of Dairy Science* **72**, 746–766.

Folman, Y., Rosenberg, M., Herz, Z. and Davidson, M. (1973). The relationship between plasma progesterone concentration and conception in post-partum dairy cows maintained on two levels of nutrition. *Journal of Reproduction and Fertility*, **34**, 267–278.

Fraser, D.L., Ørskov, E.R., Whitelaw, F.G. and Franklin, M.F. (1991). Limiting amino acids in dairy cows given casein as the sole source of protein. *Livestock Production Science* **28**, 235–252.

Fuller, M.F. (1994) Amino acid requirements for maintenance, body protein accretion and reproduction in pigs. *In Amino Acids in Farm Animal Nutrition* pp 155–184. Edited by J.P.F. D'Mello. CABI, Wallingford.

Girdler, C.P., Thomas, P.C. and Chamberlain, D.G. (1988a). Effect of intraabomasal infusions of amino acids or of a mixed animal protein source on milk production in the dairy cow. *Proceedings of the Nutrition Society* **47**, 50A.

Girdler, C.P., Thomas, P.C. and Chamberlain, D.G. (1988b). Effect of rumen-protected methionine and lysine on milk production from cows given grass silage diets. *Proceedings of the Nutrition Society* **47**, 82A.

Gordon, F. J. (1980). The effect of silage type on the performance of lactating cows and the response to high levels of protein in the supplement. *Animal Production* **30**, 29–37.

Grummer, R. (1996). Nutrition and physiology of the transition cow. *Proceedings of California Animal Nutrition Conference,* Fresno, 177–187.

Guillaume, B., Otterby, D.E., Stern, M.O., Linn, J.G. and Johnson, D.G. (1991). Raw or extruded soybeans and rumen-protected methionine and lysine in alfalfa based diets for dairy cows. *Journal of Dairy Science* **74**, 1912–1922.

Henry, Y., (1993). Affinement du concept de la protéine idéale pour le porc en croissance. *INRA Production Anim*ale, **6**, 199–212.

Huhtanen, P., Vanhartalo, A. and Varvikko, T. (1996). New knowledge about amino acid requirements of ruminants. *Rehuraisio Environmental Feed Symposium.*

Hurtaud, C., Rulquin, H. and Vérité, R. (1995). Effect of rumen protected methionine and lysine on milk composition and on cheese yielding capacity. *Annales Zootechnie* **44**, 382.

Karunanandaa, K., Goodling, L.E., Varga, G.A., Muller, L.O., McNeil, W.W., Cassidy, T.W. and Lykos, T. (1993). Supplementary dietary fat and ruminally protected amino acids for lactating Jersey cows. *Journal of Dairy Science* **77**, 3417–3425.

King, K.J., Bergen, W.G., Sniffen, C.J., Grant, A.L., Grieve, D.B., King, V.L. and Ames, N.K (1991). An assessment of absorbable lysine requirements in lactating cows. *Journal of Dairy Science* **74**, 2530–2539.

Koch, K.L., Whitehouse, N.L., Garthwaite, B.D., Wasserstom, V.M. and Schwab, C. G. (1996). Production responses of lactating Holstein cows to rumen-stable forms of lysine and methionine. *Journal of Dairy Science* **79**, (Suppl.1) 24.

Leclercq, B., (1996). Les rejets azotés issus de l'aviculture : importance et progrès envisageables. *INRA Production Animale* **9**, 91–101.

Le Henaff, L., (1991). Importance des acides aminés dans la nutrition des vaches laitières. Diplôme de doctorat. Thesis N° 253. *Université de Rennes*, France.

Loerch,S.C. and Oke, O. (1989). Rumen protected amino acids in ruminant nutrition. *In Absorption and Utilisation of Amino Acids*. Ed. Mendel Friedman Boca Raton, Florida Vol. III, 187–200.

Moorby, J.M., Dewhurst, R.J. and Mardsen, S. (1996). Effect of increasing digestible undegraded protein supply to dairy cows in late gestation on the yield and composition of milk during the subsequent lactation. *Animal Science* **63**, 201–213

O'Connor, J.D., Sniffen, C.J., Fox, D.G. and Chalupa, W. (1993). A net carbohydrate and protein system for evaluating cattle diets : IV Predicting amino acid adequacy. *Journal of Dairy Sci*ence **71**, 1298.

Ørskov, E.R., Grubb, D.A. and Kay, R.N.B. (1977). Effect of postruminal glucose or protein supplementation on milk yield and composition in Frisian cows in early lactation and negative energy balance. *British Journal of Nutrition* **38**, 397–405.

Piepenbrink, M.S., Overton, T.R. and Clark, J.H. (1996). Response of cows fed a low crude protein diet to ruminally protected methionine and lysine. *Journal of Dairy Science* **79**, 1638–1646.

Pisulewski, P.M., Rulquin, H., Peyraud, J.L. and Vérité, R. (1996). Lactational and systemic responses of dairy cows to post-ruminal infusions of increasing amounts of methionine. *Journal of Dairy Science* **79**, 1781–1791.

Piva, G., Masoero, F., Mancini, V. and Fusconi, G. (1996). The effect of the protected methionine and lysine suplementation on the performance of dairy cows in early lactation fed a low protein diet. *Journal of Animal Science* **74**, (Suppl.1) 278.

Polan, C.E., Cummins, K.A., Sniffen, C.J., Muscato, T.V., Vicini, J.L., Crooker, B.A., Clark, J.H., Johnson, D.G., Otterby, D.E., Guillaume, B., Muller, L.O., Varga, G.A., Murray, R.A. and Peirce-Sandner, S. (1991). Responses of dairy cows to supplement rumen-protected forms of methionine and lysine. *Journal of Dairy Science* **74**, 2997–3013.

Remond, B. (1988). Effet de l'addition de méthionine protégée à la ration des vaches laitières : influence du niveau des apports azotés. *Annales de Zootech*nie **37**, 271–284.

Robert, J.C. and Lavedrine, F. (1995). The effect of supplementation of hay plus soyabean meal diet with rumen protected methionine on the lactational performance of dairy cows in mid late lactation. *VII Symposium on protein metabolism and nutrition.* Portugal.

Robert, J.C., Sloan, B.K. and Bourdeau, S. (1994). The effect of supplementation of corn silage pius soyabean meal diets with rumen protected methionine on the lactational performance of dairy cows in early lactation. *Journal of Dairy Science* **77** (Suppl. 1), 349.

Robert, J.C., Sloan, B.K. and Denis, C. (1994). The effect of protected amino acid supplementation on the performance of dairy cows receiving grass silage plus soybean meal. *Animal Production* **58**, 437.

Robert, J.C., Sloan, B.K. and Denis, C. (1996). The effect of graded amounts of rumen protected methionine on lactational responses in dairy cows. *Journal of Dairy Science* **79** (Suppl. 1), 256.

Robert, J.C., Sloan, B.K. and Lahaye, F. (1995). Influence of increasing doses of intestinal digestible methionine (MetDI) on the performance of dairy cows in mid and late lactation. *IVth International Symposium on the Nutrition of Herbivores, Clermont-Ferrand,* France.

Robert, J.C., Sloan, B.K. and Nozière, P. (1996). Effects of graded levels of rumen protected lysine on lactational performance in dairy cows. *Journal of Dairy Science* **79**, (Suppl. 1) 257.

Robert, J.C., Sloan, B.K., Saby, B., Mathé, J., Dumont, G., Duron, M. and Dzyzcko, E., (1989). Influence of dietary nitrogen content and inclusion of rumen-protected methionine and lysine on nitrogen utilisation in the early lactation dairy cow. *Australasian Journal of Animal Science* **2**, 544–545.

Robert, J.C., Tanan, K., Blanchart, G., Williams, P. and Philippeau, C. (1994). Influence of the ration composition on the duodenal flow of amino acids in lactating dairy cows. *Procedings Society of Nutrition and Physiology* **3**, 66.

Robinson, P.H., Fredeen, A.H., Chalupa, W., Julien, W.E., Sato, H., Fujieda, T. and Suzuki, H. (1995). Ruminally protected lysine and methionine for lactating dairy cows fed a diet designed to meet requirements for microbial and post-ruminal protein. *Journal of Dairy Science* **78**, 582–594.

Rogers, J.A, Peirce-Sandner, S.B., Papas, A.M., Polan, C.E., Sniffen, C.J., Muscato, T.V., Staples, C.R. and Clark, J.H. (1989). Production responses of dairy cows fed various amounts of rumen-protected methionine and lysine. *Journal of Dairy Science* **72**, 1800–1817.

Rulquin, H. (1986). Influence de l'équilibre en acides aminés de trois proteines infusées dans l'intestin grêle sur la production laitière de la vache. *Reproduction Nutrition Développement* **26**, 347–348.

Rulquin, H. (1992). Interets et limites d'un apport de méthionine et de lysine dans l'alimentation des vaches laitières. *Production Animale* **5**, 29.

Rulquin, H. and Delaby, L. (1994a). Lactational responses of dairy cows to graded amounts of rumen-protected methionine. *Journal of Dairy Science* **77** (Suppl. 1), 345.

Rulquin, H. and Delaby, L. (1994b). Effects of energy status on lactational responses of dairy cows to rumen-protected methionine. *Journal of Dairy Science* **77** (Suppl. 1), 346.

Rulquin, H., Guinard, R., Vérité, R. and Delaby, L. (1993). Teneurs en Lysine (LysDI) et Méthionine (MetDI) digestibles des aliments pour ruminants. *Journées AFTAA Tours*.

Rulquin, H., Le Henaff, L. and Vérité, R. (1990). Effects on milk protein yield of graded levels of lysine infused into the duodenum of dairy cows fed diets with two levels of protein. *Reproduction Nutrition Development* (Suppl. 2) : 238.

Rulquin, H., Pisulewski, P.M., Vérité, R. and Guinard, J. (1993). Milk production and composition as a function of post-ruminal lysine and methionine supply : a nutrient response approach. *Livestock Production Science* **37**, 69.

Rulquin, H., and Vérité, R. (1993). Amino acid nutrition of dairy cows: Productive effects and animal requirements. In *Recent Advances in Animal Nutrition*, pp. 55–77. Edited by P.C. Garnsworthy an D.J.A. Cole. Nottingham University Press, Nottingham.

Schwab, C.G., Bozak, C.K., Whitehouse, N.L. and Olson, V.M. (1992). Amino acid limitation and flow to the duodenum at four stages of lactation. II. Extent of lysine limitation. *Journal of Dairy Science* **75**, 3503.

Schwab, C.G., Satter, L.D. and Clay, A. B. (1976). Response of lactating dairy cows to abomasal infusion of amino acids. *Journal of Dairy Science* **59**, 1254–1269.

Sloan, B. (1993). Ensilage de maïs/soja : quels acides aminés faut-il apporter pour augmenter le taux protéique chez la vache laitière ? *Journées AFTAA Tours.*

Sloan, B.K., Robert, J.C. and Lavedrine, F. (1994). The effect of protected methionine and lysine supplementation on the performance of dairy cows in mid lactation. *Journal of Dairy Science* 77 (Suppl.1), 343.

Sloan, B.K., Robert, J.C. and Mathé, J. (1989). Influence of dietary crude protein content plus or minus inclusion of rumen protected amino acids (RAA) on the early lactation performance of heifers. *Journal of Dairy Science* 72 (Suppl. 1), 506.

Socha, M.T. and Schwab, C.G. (1994). Developing dose response relationships for absorbable lysine and methionine supplies in relation to milk and milk protein production from published data using the Cornell Net Carbohydrate and Protein System. *Journal of Dairy Science* 77 (Suppl. 1), 93.

Socha, M.T., Schwab, C.G., Putnam, D.E., Whitehouse, N.L., Kierstead, N.A. and Garthwaite, B.D. (1994a). Determining methionine requirements of dairy cows during peak lactation by post-ruminally infusing incremental amounts of methionine. *Journal of Dairy Science* 77 (Suppl.1), 350.

Socha, M.T., Schwab, C.G., Putnam, D.E., Whitehouse, N.L., Kierstead, N.A. and Garthwaite, B.D. (1994b). Determining methionine requirements of dairy cows during early lactation by post-ruminally infusing incremental amounts of methionine. *Journal of Dairy Science* 77, 246.

Socha, M.T., Schwab, C.G., Putnam, D.E., Whitehouse, N.L., Kierstead, N.A. and Garthwaite, B.D. (1994c). Determining methionine requirements of dairy cows during mid lactation by post-ruminally infusing incremental amounts of methionine. *Journal of Dairy Science* 77 (Suppl.1), 351.

Socha, M.T., Schwab, C.G., Putnam, D.E., Whitehouse, N.L., Kierstead, N.A. and Garthwaite, B.D. (1994d). Production responses of early lactation cows fed rumen-stable methionine or rumen-stable lysine plus methionine at two levels of dietary crude protein. *Journal of Dairy Science* 77 (Suppl. 1), 352.

Thiaucourt, L. (1996). L'opportunité de la méthionine protégée en production laitière. *Bulletin des GTV* 2B, 45–52.

Thomas, C., Crocker, A., Fisher, W., Walker, C. and Reeve, A. (1989). The interaction between protein level and the response to specific nutrients in high silage diets. *Animal Production* 48, (3) 623.

Vérité, R. (1997). Developments in the INRA feeding system for dairy cows. *In Recent Advances in Animal Nutrition*, p 153–166. Edited by P.C. Garnsworthy and J. Wiseman. Nottingham University Press, Nottingham.

Whitelaw, F.G., Milne, J.S., Ørskov, E.R. and Smith, J.S. (1986). The nitrogen and energy metabolism of lactating cows given abomasal infusions of casein. *British Journal of Nutrition* 55, 537–556.

Wright, T.C., Moscardini, S., Susmel, P. and McBride, B.W. (1996). Effects of supplying graded levels of a fixed essential amino acid pattern on milk protein production and nitrogen utilisation in the lactating dairy cow. *Dairy Research Report*. University of Guelph - Publication N° 0396.

Younge, B.A., Murphy, J.J., Rath, M. and Sloan, B.K. (1995). The effect of protected methionine and lysine on milk production and composition on grass silage based diets. *Animal Science* **69**, 556.

LIST OF PARTICIPANTS

The thirty-first Feed Manufacturers Conference was organised by the following committee:

Dr M.R. Bedford (Finnfeeds International)
Dr C. Brenninkmeijer (Hendrix' Voeders Bv)
Dr W.H. Close (Close Consultancy)
Dr D.J.A. Cole (Nottingham Nutrition International)
Dr S. Jagger (Dalgety Agriculture)
Dr J.R. Newbold (BOCM Pauls Ltd.)
Dr J. O'Grady (IAWS Group Plc)
Mr P. Poornan (Lys Mill Ltd.)
Mr J.R. Pickford
Mr P.G. Spencer (Bernard Matthews Plc)
Mr D.H. Thompson (Rightfeeds Ltd)
Mr J. Twigge (Trouw Nutrition)
Dr K.N. Boorman
Prof P.J. Buttery
Dr J.M. Dawson
Dr P.C. Garnsworthy (Secretary)   } University of Nottingham
Dr W. Haresign (Chairman)
Prof G.E. Lamming
Dr A.M. Salter
Dr J. Wiseman (Chairman elect)

The conference was held at the University of Nottingham, Sutton Bonington Campus, 2nd-3rd January 1997 and the committee would like to thank the authors for their valuable contributions. The following persons registered for the meeting:

Adams, Dr C L	Kemin Europa NV, Industriezone Wolfstee, 2200 Herentals, Belgium
Adesogan, Dr A	WIRS, Llanbadarn Campus, University of Wales, Aberystwyth, SY23 3AL
Agnew, Dr R	Agricultural Research Inst NI, Hillsborough, Co Down, Northern Ireland
Allder, Mr M J	Eurotec Nutrition Ltd, Glendale House, 5B Martins Lane, Witcham, Ely, Cambs, CB6 2LB
Allen, Dr J D	Frank Wright Ltd, Blenheim House, Blenheim Rd, Ashbourne, DE6 1HA
Allen, Mrs D	Genus Management, 46A Mudford Rd, Yeovil, Somerset BA21 4AB
Allison, Mr R.	University of Nottingham, Sutton Bonington Campus, Loughborough, Leics LE12 5RD
Anderson, Mr K	Duffield Nutrition, Saxlingham Thorpe Mills, Norwich, NR15 1TY
Angold, Mr M	Roche Products Ltd, Heanor Gate, Heanor, Derbyshire, DE75 7SG
Aspland, Mr F P	Aspland & James Ltd, Medcalfe Way, Bridge St, Chatteris, Cambs
Atherton, Dr D	Thompson & Joseph Ltd, 119 Plumstead Rd, Norwich, NR1 4JT

Baker, Mr S J L	Roche Products, Heanorgate, Heanor, Derby, DE75 7SG
Ball, Mr J A	Roche Products, Heanorgate, Heanor, Derby, DE75 7SG
Barrie, Mr M	Elanco Animal Health, Kingsclere Road, Basingstoke, Hampshire, RG21 6XA
Bartram, Dr C	Dalgety Agriculture Ltd, 180 Aztec West, Almondsbury, Bristol, BS12 4TH
Bates, Mrs A	Chapman Vitrition Ltd, Ryhall Rd, Stamford, PE9 1TZ
Beard, Mr M.	University of Nottingham, Sutton Bonington Campus, Loughborough, Leics LE12 5RD
Beardsworth, Dr P M	Roche Products Ltd, Heanorgate, Heanor, Derbyshire, DE75 7SE
Beaumont, Mr D	Laboratories Pancosma, Crompton Road Ind Est, Ilkeston, Derbyshire
Beckerton, Dr A	Roche Products, Heanorgate, Heanor, Derby, DE75 7SG
Bedford, Dr M	Finnfeeds International, Box 777 Marlborough, Wilts, SN8 1XN
Beer, Mr J H	W & J Pye Ltd, Fleet Square, Lancaster, LA1 1HA
Beesty, Mr C	Lloyds Animal Feeds, Murton, Oswestry, Shropshire
Bell, Miss J F	W & J Pye Ltd, Fleet Square, Lancaster, LA1 1HA
Blake, Dr J	Dynamic Nutrition Services, Highfield, Little London, Andover, Hants
Boorman, Dr K N	University of Nottingham, Sutton Bonington Campus, Loughborough, Leics, LE12 5RD
Booth, Mrs A	Fishers Feeds, Cranswick, Driffield, Yorkshire, YO25 9PF
Borgida, Mr L P	COFNA, 25 rue du Rempart, 37018 Tours Cedex, France
Bourne, Mr S	Alltech Ltd, 16/17 Abenbury Way, Wrexham Ind Estate, Wrexham, Clwyd LL13 9UZ
Bremmers, Mr R	Loders Croklaan, P.O. Box 4, 1520 AA Wormerveer, The Netherlands
Brennan, Mr E	Red Mills, Goresbridge, Co Kilkenny, Ireland
Brenninkmeijer, Dr C	Hendrix Voeders B V, P O Box 1, 5830 MA Boxmeer, The Netherlands
Brewster, Mrs A	Fishers Feeds Ltd, Cranswick, Driffield, Yorkshire, YO25 9PF
Brooking, Ms P	Hays Ingredients, Old Gorsey Lane, Wallasey, Merseyside, L44 4AH
Brophy, Mr A	Alltech Ireland, 28 Cookstown Ind Estate, Tallaght, Dublin 24 Ireland
Brown, Mr G	Roche Products Ltd, Heanorgate, Heanor, Derbys, DE75 7SG
Brown, Mr J M	Britphos Ltd, Rawden House, Green Lane, Yeadon, Leeds LS19 7BY
Bruce, Dr D W	John Thompson & Sons Ltd, 35 York Road, Belfast, BT15 3GW
Burke, Mrs M	Nutec, Greenhills Road, Tallaght, Dublin 24, Ireland
Burnside, Mr T	British United Turkeys Ltd, Hockenhull Hall, Tarvin, Chester, CH3 8LE
Burt, Dr A W	A Burt Research Ltd, 23 Stow Road, Kimbolton, Huntingdon, Cambs, PE18 0HU
Buss, Miss J	Farmers Weekly

Buttery, Prof P J	University of Nottingham, Sutton Bonington Campus, Loughborough, Leics, LE12 5RD
Cameron, Miss J.	University of Nottingham, Sutton Bonington Campus, Loughborough, Leics LE12 5RD
Campani, Dr I	F lli Martini & C SpA, Ehilia Street 2614, 47020 Budrio di Longiano-Fo, Italy
Carter, Mr T	Anitox Ltd, Anitox House, 80 Main Road, Earls Barton NN6 OHJ
Caygill, Dr J	MAFF, Nobel House, 17 Smith Square, London, SW13 3JR
Ceccaroni, Dr F	Mangimificio Romagnolo, Via Settecrociari, 47020 S Vittore di Cesena (Fo), Italy
Charles, Dr D	David Charles R & D, 62 Main Street, Willoughby, Loughborough, LE12 6SZ
Charlton, Mr P	Alltech (UK) Ltd, 16/17 Abenbury Way, Wrexham Ind Est, Wrexham, Clwyd, LL13 9UZ
Charlton, Mrs S	Trouw Nutrition, Wincham, Northwich, Cheshire, CW9 6DF
Clarke, Mr A N	Britphos Ltd, Rawdon House, Green Lane, Yeadon, Leeds LS19 7BY
Clay, Mr J	Alltech (UK) Ltd, 16/17 Abenbury Way, Wrexham Ind Est, Wrexham, Clwyd LL13 9UX
Close, Dr W	Close Consultancy, 129 Barkham Rd, Wokingham, Berkshire, RG11 2RS
Cole, Dr D J A	Nottingham Nutrition International, 14 Potters Lane, East Leake, Loughborough, Leics, LE12 6NQ
Cole, Dr M	SCA Nutrition Ltd, Maple Mill, Dalton Airfield Ind Est, Dalton, Thirsk, N Yorks, YO7 3HE
Collyer, Mr M F	Kemin () Ltd, Becor House, Green Lane, Lincoln, LN6 7DL
Cook, Miss J Dalgety	Agriculture Ltd, 180 Aztec West, Almondsbury, Bristol, BS12 4TH
Cooke, Dr B C	Dalgety Agriculture Ltd, 180 Aztec West, Almondsbury, Bristol, BS12 4TH
Cooper, Dr A	Seale Hayne, Univ of Plymouth, Newton Abbot, , TQ12 6NQ
Coppock, Dr C	Coppock Nutritional Services, 902 Dellwood Drive, Laredo, Texas, 78041-2119 USA
Corless, Mr J	Trouw Nutrition, 36 Ship Street, Belfast, BT15 1JL Northern Ireland
Cottrill, Dr B	ADAS, Wergs Road, Wolverhampton, WV6 8TQ
Cox, Mr N	SCA Nutrition, Maple Mill, Dalton Airfield Ind Est, Thirsk, N Yorks, YO7 3HE
Creasey, Mrs A	BASF plc, Earl Rd, Cheadle Hulme, Cheshire, SK8 6QG
Davis, Dr R E	Bernard Matthews plc, Great Witchingham Hall, Norwich, Norfolk, NR9 5QD
Dawson, Dr J M	University of Nottingham, Sutton Bonington Campus, Loughborough, Leics LE12 5RD
Dawson, Mr W	Britphos Ltd, Rawdon House, Green Lane, Yeadon, Leeds LS19 7BY

De Smet, Mr P	Alltech Netherlands, Hollandsch Diep 63, 2904 EP Capelle Aanden Ijssel, The Netherlands
Dean, Mr R	Cargill Technical Services, Artioma St 1/5, Kiev, Ukraine
Deaville, Mr S	Rumenco, Stretton House, Derby Road, Stretton, Burton-on-Trent, Staffs, DE13 ODW
Dewhurst, Dr R J	IGER, Plas Gogerddan, Aberystwyth, Ceredigion, SY23 3EB
Dickerson, Mr C	Roche Products, Heanor Gate, Heanor, Derbyshire, DE75 7SG
Diepenbroek, Mr L	Mole Valley Farmers, Station Road, South Molton, Devon, EX36 3BH
Dixon, Mr D H	Brown & Gilmer Ltd, P O Box 3154, The Lodge, Florence House, 199 The Strand, Merrion, Dublin 4 Ireland
Doran, Mr B	Trouw (UK) Ltd, Wincham, Northwich, Cheshire, CW9 6DF
Drakley, Miss C.	University of Nottingham, Sutton Bonington Campus, Loughborough, Leics LE12 5RD
Drouget, Mr L	Sanders Aliments, 17 quai de l'indutrie, 91 200 Athis-mons 5005, France
Dupire, Mr C	COFNA, 25 rue du Rempart, 37018 Tours Cedex, France
Everington, Mrs J	Beacon Research Ltd, Greenleigh, Kelmarsh Road, Clipston, Leics, LE16 9RX
Ewing, Dr W	Cargill plc, Camp Road, Swinderby, Lincs, LN6 97N
Ewing, Mrs A	Dalgety Agriculture Ltd, 180 Aztec West, Almondsbury, Bristol, BS12 4TH
Filmer, Mr D	David Filmer Limited, Wascelyn, Brent Knoll, Somerset, TA9 4DT
Fitt, Dr T J	Roche Products Ltd, Heanor Gate, Heanor, Derbyshire, DE75 7SG
Fordyce, Mr J	W.M.F Ltd, Bradford Road, Melksham, Wilts, SN12 8LQ
Foulds, Mr S	Park Tonks Ltd, 48 North Road, Great Abington, Cambridge, CB1 6AS
Fulford, Mr G W	Heygates & Son Ltd, Bugbrooke Mills, Northampton
Fullarton, Mr P J	Forum Products Ltd, 41-51 Brighton Road, Redhill, Surrey, RH1 6YS
Garnsworthy, Dr P.C.	University of Nottingham, Sutton Bonington Campus, Loughborough, Leics, LE12 5RD
Geary, Mr B	Pfizer Ltd, Sandwich, Kent, CT13 9NJ
Geaves, Mr J E	Chapman Vitrition Ltd, Ryhall Rd, Stamford, Lincs
Gibson, Mr J.E.	Parnutt Foods Ltd, Hadley Rd, Woodbridge Industrial Estate, Sleaford, Lincs, NG34 7EG
Gilbert, Mr R	Asbury Publications Ltd, Stoke Road, Bishop's Cleave, Glos, GL52 4RW
Gill, Dr B P	M.L.C., PO Box 444, Winterhill House, Snowdon Drive, Milton Keynes, MK6 1AX
Gillespie, Miss F	United Molasses, Stretton House, Derby Rd, Burton on Trent, Staffs, DE13 ODW
Glover, Mr J	Elanco Animal Health, Kingsclere Road, Basingstoke, Hampshire, RG21 6XA

Golds, Mrs S.P.	University of Nottingham, Sutton Bonington Campus, Loughborough, Leics LE12 5RD
Goldsborough, Mr T	Daylay Foods Ltd, The Mill, Seamer, Stokesley
Gooderham, Mr B J	Pye Milk Products, Lansil Way, Lancaster, LA1 3QY
Goransson, Dr L A T	Swedish Pig Center, Lantmannen Foderulveckl, P 1280 26890 Svalov, Sweden
Gould, Mrs M P	Volac International Ltd, Orwell, Royston, Herts, SG8 5QX
Grace, Mr J	Elanco Animal Health, Kingsclere Road, Basingstoke, Hampshire, RG21 6XA
Gray, Mr W	Kemira Chemicals (UK) Ltd, Orm House, 2 Hookstone Park, Harrogate, HG2 7DB
Green, Dr S	Rhone Poulenc Animal Nutrition, Oak House, Reeds Crescent, Watford, WD1 1QH
Green, Miss AA	Tithebarn Ltd, P O Box 20, Tithebarn House, Weld Rd, Southport, Merseyside
Griffiths, Mr D	M S F Ltd, Defford Mill, Earls Cromme, Worcester
Hardman, Mr W	AF plc "Kinross", New Hall Lane, Preston, Lancs
Haresign, Prof W	University of Wales, Aberystwyth
Harland, Dr J	Dalgety Agriculture Ltd, 180 Aztec West, Almondsbury, Bristol, BS12 4TH
Harrison, Mrs J	Sciantec Analytical Services Ltd, Mainsite, Dalton, Thirsk, N Yorks, YO7 3JA
Haythornthwaite, Mr A	Farmsense Ltd, Wild Goose House, Goe Lane, Freckleton, PR4 1XH
Hazzledine, Dr M	Dalgety Agriculture Ltd, 180 Aztec West, Almondsbury, Bristol, BS12 4TH
Higginbotham, Dr J D	United Molasses, Avonmouth Old Dock, Avonmouth, Bristol, BS11 9BU
Hockey, Mr R A	Pfizer Ltd, 39 The Warren, Chesham, Bucks, HP5 2RX
Holder, Mr P	Intermol, Shell Road, Royal Edward Dock, Avonmouth, Bristol, BS11 9BW
Holma, Mrs M	Raisio Feed Ltd, P O Box 101, FIN-21201, Raisio, Finland
Hotten, Dr P M	Rowett Research Services, Greenburn Road, Bucksburn, Aberdeen, AB21 9SB
Howie, Mr A	Nutrition Trading, Orchard House, Manor Drive, Morton Bagot, Studley, Warks B80 7ED UL
Hughes, Mr D P	NWF Agriculture Ltd, Wardle, Nantwich, CW56 AQ
Ingham, Mr R W	Kemin (UK) Ltd, Becor House, Green Lane, Lincoln, LN6 7DL
Jacklin, Mr D	Keenan TMR Centre, NAC Stoneleigh Park, Kenilworth, Warwicks, CV8 2RL
Jackson, Mr J	Nutec Ltd, Eastern Ave, Lichfield, Staffs
Jagger, Dr S	Dalgety Agriculture Ltd, 180 Aztec West, Almondsbury, Bristol, BS12 4TH

Jardine, Mr G	Guttridge Milling, 1 Mount Terrace, York, YO2 4AR
Johnson, Miss S	Kemira Chemicals Ltd, Orm House, 2 Hookstone Park, Harrogate, North Yorkshire
Jones, Mr H	Heygates & Sons Ltd, Bugbrooke Mills, Northampton
Jones, Dr E	Nutec Ltd, Eastern Avenue, Lichfield, Staffs
Keeling, Mrs S	Nottingham University Press, Unit 2, Manor Farm, Thrumpton, Nottingham NG11 0AX
Kenyon, Mr S	Alltech (UK) Ltd, 16/17 Abenbury Way, Wrexham Ind Est, Wrexham, Clwyd LL13 9UZ
Ketelaar, IR G G	Pricor B U, P O Box 51, 3420 DB Oudewater, Netherlands
Keys, Mr J	32 Holbrook Road, Stratford Upon Avon, Warwickshire, CV37 9DZ
Kies, Mr A.	Gist-brocades NV, FSI AJG, P O Box 1 2600 MA Delft, Netherlands
Knight, Dr R	Trouw Nutrition, Wincham, Northwich, Cheshire
Kuperus, Mr W	ACM, P O Box 1033, 7940 Ka Meppel, Holland
Lamming, Prof G E	University of Nottingham, Sutton Bonington Campus, Loughborough, Leics LE12 5RD
Langer, Dr S	Ralston Purina Europe Inc, 1 Place Charles de Gaule, B P 301, 78054 Saint Quentin en, Yvelines Cedex France
Law, Mr J R	Sheldon Jones Agriculture, 1st Avenue, Royal Portbury Dock, Portbury, Bristol
Lilburn, Dr M	Ohio State University, USA
Lima, Mr S	Felleskjopet Rogaland Agder, P O Box 208, 4001 Stavanger, Norway
Lowe, Dr R A	Frank Wright Ltd, Blenheim House, Blenheim Rd, Ashbourne, DE6 1HA
Lowe, Mr J	Gilbertson & Page, P O Box 321, Welwyn Garden City, Herts, AL7 1LF
Lucey, Mr P	Dairy Gold Co-op Ltd, Lombardstown, Co Cork, Ireland
Lukehurst, Mr T	Univar plc, Priestley Road, Basingstoke
Lynn, Mr N J	Cobb Breeding Company, East Hanningfield, Chelmsford, CM3 8BY
Lyons, Dr T P	Altech Inc, Biotechnology Center, 3031 Catniphill Pike, Nicholasville, KY 40356 USA
Mafo, Mr A	High Peak Feeds, Proctors (Bakewell) Ltd, 12 Ashbourne Road, Derby, DE22 3AA
Malandra, Dr F	Sildamin, Sostegno di Spessa, 27010 (Pavias), Italy
Manning, Mrs J	Town & Country Public Relations, Cornerstone House, Stafford Park 13, Telford, TF3 3OZ
Mansbridge, Miss R J	ADAS Bridgets, Martyr Worthy, Winchester, Hants
Marsden, Dr M	J Bibby Agriculture, ABN House, P O Box 250, Oundle Rd, Woodston, Peterborough, PE2 9QF
Marsden, Dr S	Dalgety Agriculture Ltd, 180 Aztec West, Almondsbury, Bristol, BS12 4YH

Marsh, Mr S	Rumenco, Stretton House, Derby Road, Stretton, Burton on Trent, Staffs DE13 ODW
Martyn, Mr S	International Additives, Old Gorsey Lane, Wallasey, Merseyside, L44 4AH
Mason, Mrs M	ICI Nutrition, Alexander House, Crown Gate, Runcorn
McCracken, Dr K J	DANI, Newforge Lane, Belfast, BT9 5PX Northern Ireland
McDonald, Mr P	David Moore (Flavours) Limited, 15/17 Abenbury Way, Wrexham Ind Est, Wrexham, LL13 9UZ
McIlmoyle, Dr A	Animal Nutrition & Agric Consultants, 20 Young St, Lisburn, BT27 5EB N Ireland
McLean, Mr R	W L Duffield & Sons Ltd, Saxlingham Thorpe Mills, Norwich, NR15 1TY
Mills, Mr C.	University of Nottingham, Sutton Bonington Campus, Loughborough, Leics LE12 5RD
Mors, Mr R	Avebe, Avebeweg 1, 9607 PT Foxhole, The Netherlands
Moss, Mrs A	ADAS Feed Evaluation Unit, Drayton Manor Drive, Alcester Rd, Stratford upon Avon CV37 9RQ
Mounsey, Mr S P	HGM Publications, Abney House, Baslow, Bakewell, Derbyshire, DE45 1RZ
Mounsey, Mr A D	HGM Publications, Abney House, Baslow, Bakewell, Derbyshire, DE45 1RZ
Mounsey, Mr H	HGM Publications, Abney House, Baslow, Bakewell, Derbyshire, DE45 1RZ
Mudd, Mr A J	Roche Products Ltd, Heanor Gate, Heanor, Derby, DE75 7SG
Murray, Mr F	Dairy Crest Ingredients, Dairy Crest House, Portsmouth Road, Surbiton, Surrey
Naish, Sir D	National Farmers Union, Agriculture House, 164 Shaftesbury Ave, London, WC2H 8HL
Neelson, Mr T	Alltech (Netherlands), Hollandsch Diep 63, 2904 EP Capelle AAnden Ijssel, Netherlands
Newbold, Dr J R	BOCM Pauls Ltd, 47 Key Street, Ipswich
Nolan, Mr J	International Additives, Old Gorsey Lane, Wallasey, Merseyside, L44 4AH
Nordang, Dr L	Felleskjopet Forutvikling, N-7005 Tronheim, , Norway
O'Grady, Dr J	IAWS Group plc, 151 Thomas St, Dublin 8, Ireland
Oram, Mr D	Roche Products Ltd, Heanor Gate, Heanor, Derbyshire, DE75 7SG
Overbeek, Dr G J	Borculo Whey Products, 7270 AA Borculo, Needseweg 23, Holland
Overend, Dr M A	Nutec Ltd, Eastern Ave, Lichfield, Staffs
Packington, Mr A J	Roche Products Ltd, Heanor Gate, Heanor, Derbyshire, DE75 7SG
Pallister, Dr S M	Orffa (UK) Ltd, Tecnavet House, Park Street, Congleton, Cheshire, CW12 1ED

Partridge, Mr M	Pen Mill Feeds Ltd, Babylon View, Pen Mill Trading Estate, Yeovil, Somerset
Partridge, Dr G	Finnfeeds International, Box 777 Marlborough, Wilts
Perrott, Mr G	Trident Feeds, P O Box 11, Oundle Road, Peterborough
Perry, Mr F	Bio Ag Connections, Pencoed Farm, Pontypridd, Mid Glamorgan
Petersen, Miss S T	Queensland Poultry, R & D Centre, P O Box 327, Cleveland, QLD 4163, Australia
Phillips, Mr G	Silo Guard Europe, Greenway Farm, Cheltenham, GL52 6PL
Pickford, Mr J R	Bocking Hall, Bocking Church Street, Braintree, Essex, CM7 5JY
Pike, Dr I H	IFOMA, 2 College Yard, St Albans, AL3 4PA
Piva, Dr G	Universita Cattalica S Cuore, Via Emilia Parmense 84, 29100 Piacenza, Italy
Plowman, Mr G B	G W Plowman & Son Ltd, Selby House, High Street, Spalding, Lincs
Poornan, Mr P	Lys Mill Ltd, Watlington, Oxon, OX9 5ES
Prinslow, Mr J	Alltech (UK) Ltd, 16/17 Abenbury Way, Wrexham Ind Estate, Wrexham, Clwyd LL13 9UZ
Probert, Miss L.	University of Nottingham, Sutton Bonington Campus, Loughborough, Leics LE12 5RD
Pusztai, Dr A	Rowett Research Institute, Bucksburn, Aberdeen, AB20 9SB
Rae, Dr R C	Premier Nutrition Products Ltd, The Levels, Rugeley, Staffs, WS15 1RD
Raine, Dr H	J Bibby Agriculture, Wildmere Rd Ind Est, Wildmere Rd, Banbury, Oxon
Raper, Mr G J	Laboratories Pancosma, Crompton Road Ind Est, Ilkeston, Derbyshire
Raper, Mr M W	Pen Mill Feeds Ltd, Babylon View, Pen Mill Trading Est., Yeovil, Somerset
Reeve, Dr A	ICI Nutrition, Alexander House, Crown Gate, Runcorn
Reynolds, Dr C K	University of Reading, Department of Agriculture, Earley Gate, Reading, RG6 6AT
Rider, Miss S	Farmers Weekly
Roberts, Mr J C	Harper Adams College, Newport, Shropshire
Robertshaw, Miss K	BOCM Pauls Ltd, 47 Key Street, Ipswich
Robinson, Mr D K	Favor Parker Ltd, The Hall, Stoke Ferry, Kings Lynn, PE33 9SE
Rosen, Dr G D	66 Bathgate Road, London SW19 5PH
Salter, Dr A	University of Nottingham, Sutton Bonington Campus, Loughborough, Leics LE12 5RD
Scheele, Dr C W	ID-DLO, Runderweg 2, PO Box 65, NL-8200 AB The Netherlands
Schoterman, Mr G.	Protonweg B, 3504 AB, Utrecht, The Netherlands
Schulze, Dr H	Finnfeeds International, Box 777, Marlborough, Wilts, SN8 1XN
Sheehy, Dr T	University College Cork, Dept of Nutrition, Ireland

Shilton, Mrs L	Frank Wright, Blenheim House, Blenheim Rd, Ashbourne, DE6 1HA
Shipton, Mr P	Thomas Flynn & Sons Ltd, The Downs, Mullingar, Co West Heath, Eire
Shorrock, Dr C	FSL Bells, Hartham, Corsham, Wilts, SN13 OQB
Short, Miss F.J.	University of Nottingham, Sutton Bonington Campus, Loughborough, Leics LE12 5RD
Singer, Dr M	Lohmann Animal Health, 3 Brownfield Way, Wheathampstead, Herts, AL4 8LL
Sitwell, Miss N	Crediton Milling Co, Fordton Mills, Crediton, Devon, EX17 3DH
Sloan, Dr B	Rhone-Poulenc, 42 Avenue Aristide Briand, BP 100, 92164 Antony Cedex France
Sloan, Mr J F	Roche Products Ltd, 25 Stockmans Way, Belfast, BT9 7JX Northern Ireland
Spencer, Mr P G	Bernard Matthews plc, Great Witchingham Hall, Norwich, Norfolk, NR9 5QD
Stainsby, Mr A K	B A T A Ltd, Norton Road, Malton, N Yorks, YO17 ONU
Steinbock, Mr M	Forum Products Ltd, 41-51 Brighton Road, Redhill, Surrey, RH1 6YS
Street, Mr C	Univar Plc, Priestley Road, Basingstoke
Sutton, Dr J D	University of Reading, Department of Agriculture, Early Gate, Reading, RG6 6AT
Swarbrick, Mr J	Borculo Whey Products, Brymau Four Estate, River Lane, Saltney, Chester, CH4 8RQ
Sylvester, Mr D	Roche Products Ltd, Heanor Gate, Heanor, Derbyshire, DE75 7SG
Taylor, Dr S J	Nutec Ltd, Eastern Ave, Lichfield, Staffs
Taylor, Dr A J	Roche Products Ltd, Heanor Gate, Heanor, Derbyshire, DE75 7SG
Ten Doeschate, Dr R	Dalgety Agriculture Ltd, 180 Aztec West, Almondsbury, Bristol, BS12 4TH
Thomas, Dr C	Scottish Agricultural College, Auchincruive, Ayr, Scotland, KA6 5HW
Thompson, Miss J.	University of Nottingham, Sutton Bonington Campus, Loughborough, Leics LE12 5RD
Thompson, Mr D	Rightfeeds Ltd, Woodlawn, Castlegate, Caddamore, Co Limerick, Ireland
Thompson, Mr R	AF plc "Kinross", New Hall Lane, Preston, Lancs
Tibble, Mr S	SCA Nutrition Ltd, Maple Mill, Dalton Airfield Ind Est, Dalton, Thirsk, N Yorks, YO7 3HE
Toplis, Mr P	Primary Diets, Melmerby In Estate, Melmerby, Ripon, N Yorks, HG4 5HP
Torvi, Ms J	Stormellen, Boks 2534, 7002 Tronheim, Norway
Townson, Mr J	Alltech (UK) Ltd, 16/17 Abenbury Way, Wrexham Ind Est, Wrexham, Clwyd LL13 9UZ
Treacher, Dr R	Finnfeeds International, Box 777, Marlborough, Wilts, SN8 1XN
Trebble, Mr J W	Mole Valley Farmers, Station Rd, South Molton, Devon, EX36 3BH

Tuck, Mr K	Alltech Ireland, 28 Cookstown Ind Estate, Tallaght, Dublin 24 Ireland
Twigge, Mr J	Trouw Nutrition, Wincham, Northwich, Cheshire, CW9 6DF
Uprichard, Mr J	Trouw Nutrition, 36 Ship St, Belfast, BT15 1JL Northern Ireland
van Cauwenberghf, Ms S	Eurolysine, 16 Rue Ballu, 75 424 Paris CEDEX OG, France
Van der Velden, Mr GCTM	Orffa Nederland Feed BV, Burgstraat 12, 4283 GG Giessen, Netherlands
Van Straalen, Dr W	Institute for Animal Nutrition, "De Schothorst", Meerkoetenweg 26, P O Box 533, 8200 AM Lelystad, The Netherlands
Van der Ploeg, Mr H	Stationsweg 4, 3603 EE Maarssen, Netherlands
Vanstone, Mr M J	Crediton Milling Co Ltd, Fordton Mills, Crediton, Devon, EX17 3DH
Verite, Mr R	INRA - SRVL Rennes, 35.590 Saint Gilles, , France
Vik, Mr K R	Stormollen As, 5270 Vaksdal, Norway
Wales, Mr C M	Dalgety Agriculture, West Quay Road, Poole, Dorset
Wallace, Mr J	Nutrition Trading Int Ltd, Orchard House, Manor Drive, Morton Bagot, Studley, Warwickshire, B80 7ED
Wareham, Dr C N	Grain Harvesters Ltd, The Old Colliery, Wingham, Canterbury, Kent
Williams, Mr P G	Akzo Nobel Surface Chemistry Ltd, 23 Grosvenor Road, St Albans, Herts, AL1 3AW
Williams, Mr D J	Intermol, Shell Road, Royal Edward Dock, Avonmouth, Bristol, BS11 9BW
Willis, Mrs C	Aynsome Laboratories, Eccleston Grange, Prescot Road, St Helens, Merseyside
Wilson, Dr J	Cherry Valley Farms Ltd, N Kelsey Moor, Caistor, Lincs, LN7 6HK
Wiseman, Dr J	University of Nottingham, Sutton Bonington Campus, Loughborough, Leics LE12 5RD
Yeo, Dr G W	Premier Nutrition Products Ltd, The Levels, Rugeley, Staffs, WS15 1RD
Youdan, Dr J	Nutrimix, Boundary Ind Est, Boundary Rd, Lytham, Lancs, FY8 5HU
Zwart, Mr S	Tessenderlo Chemie Rotterdam B V, P O Box 133, 3130 AC Vlaardingen, Netherlands

INDEX